AF560390

QUANTUM CHEMISTRY

QUANTUM CHEMISTRY

Jyoti Kumar Roshan

ANMOL PUBLICATIONS PVT. LTD.
NEW DELHI - 110 002 (INDIA)

ANMOL PUBLICATIONS PVT. LTD.
H.O.: 4374/4B, Ansari Road, Darya Ganj,
New Delhi-110 002 (India)
Ph.: 23278000, 23261597
B.O.: No. 1015, Ist Main Road, BSK IIIrd Stage
IIIrd Phase, IIIrd Block,
Bangalore - 560 085 (India)
Visit us at: www.anmolpublications.com

Quantum Chemistry

ISBN 978-81-261-3652-0

PRINTED IN INDIA

Printed at Mehra Offset Press, Delhi.

Contents

Preface

Quantum chemistry has long promised to become a major tool for the study of molecular properties and reaction mechanisms. The fundamental methods of quantum chemistry date back to the earliest days of quantum mechanics in the first decades of the twentieth century. Quantum mechanics, particularly quantum chemistry, is a crucial part of the language of modern chemical science. Terms such as pð-dð interaction, symmetry-forbidden reactions, bond-order/bond-energy relationships, hypervalency, and exchange repulsion are typical of those that arise in contemporary discussions of chemical structure and properties.

This language is a powerful one with which the contemporary chemical scientist needs to be familiar. The principal aim of this book is to "bridge the gaps" between traditional quantum chemistry, which is mainly concerned with the calculation of the electronic properties of molecules, theoretical high resolution spectroscopy, traditionally concerned with the calculation of bound rovibronic states, and molecular dynamics which looks at the time dependence of molecular processes. The book is aimed at research workers in all of these fields, and in the border areas between them.

Chapter 1

Introduction

THE SCHRODINGER EQUATION

Rutherford established, that atoms are comprised of a dense, positively-charged nucleus which is surrounded by a region of negative charge. It had been established that the negatively-charged region was comprised of particles called electrons. In the period between the discovery of these basic atomic attributes and quantum theory, little was known as to why the electrons in atoms are not instantly attracted into the nucleus. Afterall, there does exist an electrostatic potential between the nucleus and the electrons. What kept them from falling inward, thereby causing the atom to collapse?

An Austrian physicist named Erwin Schrodinger confronted the problem, utilizing a new approach which was emerging at the time. He used the idea that matter in motion possesses wave-like attributes. This idea, which may seem unusual to those first exposed to it, was not based on pure speculation. In fact, much work was being produced at this point which showed that light exists as both particle and wave, a concept which originated from experiments concerning the diffraction of light around barriers and blackbody radiation. A French physicist named Louis Victor de Broglie had applied this particle-wave concept to matter and had devised an expression which relates the momentum of an object with the wavelength of its 'matterwave'.

Not only did Schrodinger utilize the idea of waves in his new theory, but also the idea of quantization. A Danish physicist named Niels Bohr had attempted to work with this idea after he realized that the emission spectrum of hydrogen only possessed certain bands of colour. These bands indicated to him that the electrons of hydrogen were making certain energy transitions within the atom, meaning the energy which an electron could possess was restricted to a discrete set of values. As it turns out, standing waves do in fact display a sort of quantization in that there can only exist integer amounts of antinodes and nodes along the wave.

Schrodinger's model for electron behaviour is usually referred to as the wave-mechanical model. This model states that all the possible positions which an electron can occupy, can be represented by a wave. It does so by ascribing a specific function to the electron. Since the function is based on wave properties, it is named a wave function. From this wave function, all information which one wishes to ascertain about the electron can be extracted. Wave functions will vary depending on the situation. The only conditions placed on them is that they must be continuous, single valued and square-integrable. The Schrodinger Equation is the mathematical relation between an electron's wavefunction and total mechanical energy. The Schrodinger Equation is a very generalized equation, making it versatile in terms of its application to various types of electron behaviour.

TIME-INDEPENDENT DERIVATION OF THE SE:

The Schrodinger Equation is based upon two formulas: the classical wave function as derived from Netwon's Second Law, and the DeBroglie wave expression. To begin deriving the time-independent form of the Schrodinger Equation, the formula for a classical wave is used first:

$$\frac{d^2Z}{dx^2} = \frac{1}{c^2}\frac{d^2z}{dt^2}$$

The letter z stands for the amplitude of the wave which is a function of x and t: z(x,t), c is the speed of the wave through the medium. The function z is broken down into two functions each being only dependent on one independent variable, either x or t.

$$Z(x,t) = \psi(x)\zeta(t)$$

The first function y represents the spatial part of the original function and is dependent only on x, the second function function ζ represents the temporal part of the original function and is dependent only on t. Using this, the following is obtained:

$$\zeta\frac{d^2\psi}{dx^2} = \frac{\psi}{c^2}\frac{d^2\zeta}{dt^2}$$

Notice only one function is differentiated in each term. Rearranging in order to separate y and z, produces:

$$\frac{1}{\psi}\frac{d^2\psi}{dx^2} = \frac{1}{c^2\zeta}\frac{d^2\zeta}{dt^2}$$

In order for this to be true, both sides of the equation must equal each other for all values of x and t. Therefore, each side can be set equal to an arbitrary constant, allowing each function to be analysed separately. In this case, the arbitrary constant utilized will be $-k^2$, which allows for the greatest possible ease when manipulating the second-order differential equations present. The left-hand side is the time-independent part and produces the time-independent form of the Schrodinger Equation. Dealing with this side first makes it easier later on to deal with the full, time-dependent form of the Schrodinger Equation. Concentrating now on the left side:

$$\frac{1}{\psi}\frac{d^2\psi}{dx^2} = -k^2$$

Rearranging:

$$\frac{d^2\psi}{dx^2} = -k^2\psi$$

The general form of y in this equation is:

$$\psi(x) = Ne^{\pm ikx}$$

where N is a normailzing constant (important in dealing with actual numerical problems). Now it is important to determine the expression to which kappa is equal so that a definite set of wave functions can be ascertained. Realizing that this function describes the amplitude of a wave, one can assume that the amplitude at one point should equal that at a different point one wavelength removed. Representing this:

$$Ne^{\pm ikx} = Ne^{\pm ik(x+\lambda)}$$

where λ is the wavelength of the wave. In order for this statement to be true, one observes:

$$k\lambda = 2\pi$$

Now the DeBroglie wave expression mentioned earlier is utilized, which is:

$$\frac{h}{mv} = \lambda$$

Substituting this into Equation, produces our representation of the values for k:

$$k = \frac{2\pi mv}{h}$$

Then, using this, one can return to equation to continue working towards the time-independent Schrodinger Equation which relates wavefunction with energy. Doing so:

$$\frac{d^2\psi}{dx^2} = -\frac{4\pi^2 m^2 v^2}{h^2}\psi$$

A convention is made by use of a symbol called h-bar and is represented as:

$$\hbar = \frac{h}{2\pi}$$

This symbol appears so often in quanutm chemistry, that this convention is an absolute necessity. Rewritting:

$$\frac{d^2\psi}{dx^2} = -\frac{m^2v^2}{\hbar^2}\psi$$

Isolating y on the right-hand side, produces:

$$\frac{-\hbar^2}{m^2v^2}\frac{d^2\psi}{dx^2} = \psi$$

Using the expression for kinetic energy,:

$$E_{kinetic} = \frac{1}{2}mv^2$$

equation can be rewritten, so that the relationship between a wavefunction and its energy is explicitly shown

$$\frac{-\hbar^2}{2m}\frac{d^2\psi}{dx^2} = E\psi$$

thereby resulting in the time-independent form of the Schrodinger Equation. This form can be modified to show realistic behaviour by adding a potential energy term on the left side as well as replacing d^2/dx^2 with the Laplacian operator (del^2) which represents taking the full second derivative of ψ:

$$\frac{-\hbar^2}{2m}\nabla^2\psi + V\psi = E\psi$$

Using operator notation, the Schrodinger Equation can be represented as:

$$\hat{H}\psi = E\psi$$

H is called a hamiltonian operator or simply the hamiltonian. A mathematical operator is a set of instructions that acts on a function, an eigenfunction, to produce a value that is characteristic of that function, an eigenvalue. In this case, the hamiltonian operator acts on ψ, the wave function, and produces a characteristic value, energy. The explicit form of

H can vary between situations. For more complex manipulations, it condenses the work expended to simply utilize H to represent several operations. Operators are demonstrated in greater detail later on.

TIME-DEPENDENT DERIVATION OF THE SE:

The time-dependent form of the Schrodinger Equation is useful for certain cases in which the state of the electron is changing as a function of time. To derivate this form, one refers back to equation of the time-independent derivation:

$$\frac{1}{\psi}\frac{d^2\psi}{dx^2} = \frac{1}{c^2\zeta}\frac{d^2\zeta}{dt^2}$$

Each side of the equation is set equal to a constant, -k ^{2}:

$$-k^2 = \frac{1}{c^2\zeta}\frac{d^2\zeta}{dt^2}$$

Multiplying by c^2 to eliminate the constants on the right-side:

$$-\beta^2 = \frac{1}{\zeta}\frac{d^2\zeta}{dt^2}$$

where $-b^2 = -k^2 * c^2$. The solutions to ζ are similar to those derived previously for the function y:

$$\zeta(t) = Ne^{\pm i\beta t}$$

where N is a normalizing constant. To determine the values of β, one must apply the principals used to determine k in the time-independent equation by recognizing this solution describes a wave, specifically a traveling wave. The amplitude located at a point at time t, should be equal to the amplitude at the same point at time t + τ, where τ is the period of the wave. Representing this:

$$Ne^{\pm i\beta t} = Ne^{\pm i\beta(t+\tau)}$$

For this equation to be true, one observes:

$$\beta\tau = 2\pi$$

Knowing $\tau = 1/v$, where v is the frequency of the wave, one can rewrite β in terms of v (this is done for future convience when solving for energy):

$$\beta = 2\pi v$$

Entering this into the general form for ζ:

$$\zeta(t) = Ne^{\pm 2\pi i v t}$$

Knowing E = h * v and h-bar=h / (2*π), one can write:

$$\zeta(t) = Ne^{-iEt/\hbar}$$

For energy to be positive, the negative sign is chosen. Earlier, the classical function z which was a function of space and time was broken into two functions. To derive the time-dependent Schrodinger Equation, one must find the quantum version of the function z which is labeled Ψ and is represented as the product of the two 'sub-functions':

$$\psi(x,t) = \psi(x)\zeta(t) = \psi e^{-iEt/\hbar}$$

To extract E from this expression and thereby arrive at the time-dependent form, the first derivative with respect to time is taken:

$$\frac{\partial\psi}{\partial t} = \frac{-iE}{\hbar}\psi e^{-iEt/\hbar}$$

Knowing –i = 1/i:

$$\frac{\partial\psi}{\partial t} = \frac{E}{i\hbar}\psi e^{-iEt/\hbar}$$

Rearranging:

$$i\hbar\frac{\partial\psi}{\partial t} = E\psi$$

finally results in the time-dependent form of the Schrodinger Equation. This can also be written as:

$$\hat{H}\psi = E\psi$$

similar to the time-independent form.

TIME-DEPENDENCE OF EIGENVALUES:

When dealing with time-dependent systems, one may want to ascertain the rate at which eigenvalues change with respect to time. Why any normal person would want to do this, one can't be sure. There is a means of determining the change in these values. To find an easy relation to work with, one begins with the definition of an eigenvalue:

$$\langle\Omega\rangle = \left\langle \psi_N \middle| \hat{\Omega} \middle| \psi_N \right\rangle = \int \psi_N^* \hat{\Omega} \psi_N d\tau$$

The sub-script N indicates a normalized wave function. Taking the time-derivative of both sides:

$$\frac{\partial\langle\Omega\rangle}{\partial t} = \frac{\partial}{\partial t} \int \psi_N^* \hat{\Omega} \psi_N d\tau$$

Applying the differential operator through, keeping in mind the change in time of the operator Ω is zero:

$$\frac{\partial\langle\Omega\rangle}{\partial t} = \int \frac{\partial \psi_N^*}{\partial t} \hat{\Omega} \psi_N d\tau + \int \psi_N^* \hat{\Omega} \frac{\partial \psi}{\partial t} d\tau$$

To get anywhere at this point, substitutions are used which originate from the time-dependent form of the Schrodinger equation:

$$\hat{H}\psi = i\hbar \frac{\partial \psi}{\partial t}, \; \hat{H}\psi^* = -i\hbar \frac{\partial \psi^*}{\partial t}$$

Using these:

$$\frac{\partial\langle\Omega\rangle}{\partial t} = \frac{-1}{i\hbar} \int \psi_N^* \hat{H} \hat{\Omega} \psi_N d\tau + \frac{1}{i\hbar} \int \psi_N^* \hat{\Omega} \hat{H} \psi_N d\tau$$

Simplifying:

$$\frac{\partial\langle\Omega\rangle}{\partial t} = \frac{-1}{i\hbar} \left[\int \psi_N^* \hat{H} \hat{\Omega} \psi_N d\tau - \int \psi_N^* \hat{\Omega} \hat{H} \psi_N d\tau \right]$$

Knowing –1 / i = i and consolidating:

$$\frac{\partial\langle\Omega\rangle}{\partial t}=\frac{-1}{i\hbar}\left\langle\psi_N\left|\hat{H}\hat{\Omega}-\hat{\Omega}\hat{H}\right|\psi_N\right\rangle$$

Rewritting the observable as a commutator produces a generalized equation for determining the rate of change of an eigenvalue with time:

$$\frac{\partial\langle\Omega\rangle}{\partial t}=\frac{i}{\hbar}\left\langle\left[\hat{H}-\hat{\Omega}\right]\right\rangle$$

Equation 8 can be applied to operators such as the linear momentum operator:

$$\frac{\partial\langle\hat{p}_x\rangle}{\partial t}=\frac{i}{\hbar}\left\langle\left[\hat{H},\hat{p}_x\right]\right\rangle$$

The linear momentum operator is of the form:

$$\hat{p}_x=\frac{\hbar}{i}\nabla$$

Using this:

$$\frac{\partial\langle\hat{p}_x\rangle}{\partial t}=\frac{i}{\hbar}\left(\left[\frac{-\hbar^2}{2m}\nabla^2+V\right]\left[\frac{\hbar}{i}\nabla\psi\right]-\frac{\hbar}{i}\nabla\left[\frac{-\hbar^2}{2m}\nabla^2\psi+V\psi\right]\right)$$

Expanding, retaining the order in which the operators act on the arbitrary function:

$$\frac{\partial\langle\hat{p}_x\rangle}{\partial t}=\left(\frac{-\hbar^2}{2m}\nabla^3\psi+V\nabla\psi-\frac{-\hbar^2}{2m}\nabla^3\psi-\nabla V\psi\right)$$

Simplifying produces the final result:

$$\frac{\partial\langle\hat{p}_x\rangle}{\partial t}=V\nabla\psi-\nabla V\psi=-\nabla V\psi$$

The negative gradient of the potential in which the electron exists is equal to the change in momentum of the electron with respect to time. This quantity represents the mean force that the electron experiences while in the potential. The change in the mean position of the electron with respect to time can also be determined through the same procedure. Starting with equation 8:

$$\frac{\partial\langle\hat{X}\rangle}{\partial t}=\frac{i}{\hbar}\langle[\hat{H},\hat{x}]\rangle$$

Writing out the commutator explicitly:

$$\frac{\partial\langle\hat{X}\rangle}{\partial t}=\frac{i}{\hbar}\left(\left[\frac{-\hbar^2}{2m}\nabla^2+V\right][x\psi]-x\left[\frac{-\hbar^2}{2m}\nabla^2\psi+V\psi\right]\right)$$

Expanding, retaining the order in which the operators act on the arbitrary function:

$$\frac{\partial\langle\hat{X}\rangle}{\partial t}=\frac{i}{\hbar}\left(\frac{-\hbar^2}{2m}\nabla^2[x\psi]+V_x\psi-\frac{-\hbar^2}{2m}x\nabla^2\psi-Vx\psi\right)$$

Simplifying:

$$\frac{\partial\langle\hat{X}\rangle}{\partial t}=\frac{i}{\hbar}\left(\frac{-\hbar^2}{2m}\left(2\nabla\psi+x\nabla^2\psi\right)-\frac{-\hbar^2}{2m}x\nabla^2\psi\right)$$

Further simplifying produces the final result:

$$\frac{\partial\langle\hat{X}\rangle}{\partial t}=\frac{\hbar}{im}\nabla\psi=\frac{\langle\hat{p}_x\rangle}{m}\psi$$

The change in the mean position of the electron with respect to time is equal to the average linear momentum divided by the mass of the electron. One can assert that this is also equal to the mean velocity of the electron which isn't all that surprising.

Probability and Current Densities:

The probability density for an infinitesimal region is equal to the square modulus of the wave function, as proposed by the physicist Max Born.

$$\rho d\tau=\psi^*\psi d\tau$$

One can utilize a form of Gauss' divergence theorm to obtain more information concerning this, which is:

$$-\nabla\vec{J} = \dot{\rho}$$

Rewriting the time derivative fo the probability density function:

$$-\nabla\vec{J} = \frac{\partial\left(\psi^{*}\psi\right)}{\partial t}$$

Expanding:

$$-\nabla\vec{J} = \frac{\partial\psi^{*}}{\partial t}\psi + \psi^{*}\frac{\partial\psi}{\partial t}$$

The time-dependent Schrodinger Equation for the free particle is of the form:

$$\frac{-\hbar^2}{2m}\nabla^2\psi = i\hbar\frac{\partial\psi}{\partial t},\ \frac{-\hbar^2}{2m}\nabla^2\psi^{*} = -i\hbar\frac{\partial\psi^{*}}{\partial t}$$

Rewriting:

$$\frac{-\hbar^2}{2mi}\nabla^2\psi = \frac{\partial\psi}{\partial t},\ \frac{\hbar^2}{2mi}\nabla^2\psi^{*} = \frac{\partial\psi^{*}}{\partial t}$$

Applying these two equations in equation:

$$-\nabla\vec{J} = \psi\left(\frac{\hbar}{2mi}\nabla^2\psi^{*}\right) + \psi^{*}\left(\frac{-\hbar}{2mi}\nabla^2\psi\right)$$

Collecting:

$$-\nabla\vec{J} = \frac{\hbar}{2mi}\left(\psi\nabla^2\psi^{*} - \psi^{*}\nabla^2\psi\right)$$

Canceling the negative sign and the operator-vector del:

$$\vec{J} = \frac{\hbar}{2mi}\left(\psi^{*}\nabla\psi - \psi\nabla\psi^{*}\right)$$

FREE-PARTICLE MODEL

The free particle model for electron behaviour is the simplest which can be considered since it places no energy restrictions on the particle. One begins with the time-independent Schrodinger Equation. Time-dependent wave functions are acceptable to use in this case since they are eigenfunctions of the time-independent hamiltonian:

$$\frac{-\hbar^2}{2m}\nabla^2\psi + V\psi = E\psi$$

Since no electrostatic potentials exist to influence the electron's behaviour, V(r) = 0. Rearranging and using a substitution, k:

$$\nabla^2\psi + k^2\psi = 0,\ k^2 = \frac{2mE}{\hbar^2}$$

Solving this homogeneous differential equation produces two linearly independent solutions to ψ:

$$\psi(\bar{r},t) = c_1 e^{i(kr-wt)} + c_2 e^{-i(kr-wt)}$$

This two-parameter solution of can be reduced to one-parameter for simplicity if we assume the electron's momentum vector is directed in only the positive r direction at any values of r and t.

$$\psi(\bar{r},t) = Ne^{i(kr-wt)}$$

Using trigonometric identities to expand this solution:

$$\psi(\bar{r},t) = N[\cos(kr) + i\sin(kr)][\cos(wt) - i\sin(wt)]$$

It's obvious that the wave function is a complex function. Complex quantities can be separated into their real and imaginary parts. The real component of ψ can be separated:

$$\operatorname{Re}\psi(\bar{r},t) = N[\cos(kr)\cos(wt) + \sin(kr)\sin(wt)]$$

Rewriting using trigonometric identities:

$$\operatorname{Re}\psi(\bar{r},t) = N\cos(kr - wt)$$

The imaginary component of ψ can be extracted as well:

$$\operatorname{Im}\psi(\bar{r},t) = N[\sin(kr)\cos(wt) - \cos(kr)\sin(wt)]$$

Rewriting using trigonometric identities:

$$\text{Im}\,\psi(\vec{r},t) = N\sin(kr - wt)$$

The Free-Particle model is not an entirely realistic representation of electron behaviour in the absence of electromagnetic fields. The reason being that the solutions for ψ listed above are not square-integrable, one of the three conditions placed on wave functions. Such solutions imply the electron probability density is uniform throughout space, which isnt a realistic description of a particle. The use of wavepackets to realistically describe the behaviour of a free electron.

THE HARMONIC OSCILLATOR MODEL

The harmonic oscillator model is important because not only does it demonstrate quantization of energy, but it also shows the phenomenon called quantum tunnelling, in which an electron can pass into a finite potential barrier, which is not permitted by classical mechanics.

This model is similar to the one-dimensional box situation, the electron is confined to a one-dimensional region, the x-axis. This case is different however, in that the electrostatic potential is different. Instead of being zero between two points and approaching infinity in the regions beyond, the potential increases with the power of x^2. The potential is centered such that its value at x=0 is zero. The electron can be thought of as being attached to one end of an imaginary spring, with the other end secured at the origin. For this system, the energy expression will be derived first, then the wave functions will be derived second (this makes it easier to understand what is going on). To begin, one first writes the Schrodinger Equation, in this case the time-independent form.

$$\frac{-\hbar^2}{2m}\frac{d^2\psi}{dx^2} + \frac{kx^2}{2}\psi = E\psi$$

Notice in this case, the potential energy term is everywhere nonzero, unlike the one-dimensional box model in which it was zero for the region of interest. The potential energy term is identical to the energy representation for a spring in classical mechanics, where k is the spring constant of the imaginary spring to which the electron is attached. To begin simplifying, in order that this expression may later be broken down, the second-order term on the left is isolated from any constants. One multiplies by 2m /–(h–bar)^2:

$$\frac{d^2\psi}{dx^2} - \frac{mkx^2}{\hbar^2}\psi = -\frac{2mE}{\hbar^2}\psi$$

Equation is a non-homogeneous differential equation with variable coefficients. Even the simplest of systems often produce complicated expressions which take time to solve, mostly by relating them with differential equations already solved, or by using approximation schemes to find a reliable answer. This equation is similar to one solved by Charles Hermite, a French mathematician. To begin solving, the asymptotic solution to this expression is found by noting which terms disappear for large values of $|x|$. Here, the term on the right-side goes to zero, relative to the terms on the left-side. Rearranging:

$$\frac{d^2\psi_\infty}{dx^2} = \frac{mkx^2}{\hbar^2}\psi_\infty$$

The asymptotic form of the wave function is labeled ψ_{Inf}. The general solution to ψ_{Inf} is:

$$\psi_\infty = \exp\left(\frac{-x^2\sqrt{mk}}{2\hbar}\right)$$

The solution to the actual wave function is the product of the asymptotic solution and another function, which is labeled here as S:

$$\psi = \psi_\infty S(x)$$

Inserting this expression into equation yields:

$$\frac{d^2\psi_\infty}{dx^2}S + 2\frac{d\psi_\infty}{dx}\frac{dS}{dx} + \psi_\infty\frac{d^2S}{dx^2} - \frac{mkx^2}{\hbar^2}\psi_\infty S = -\frac{2mE}{\hbar^2}\psi_\infty S$$

Rewritting explicitly:

$$\left[\frac{mkx^2}{\hbar^2}S-\frac{\sqrt{mk}}{\hbar}S-\frac{2x\sqrt{mk}}{\hbar}\frac{dS}{dx}+\frac{d^2S}{dx^2}-\frac{mkx^2}{\hbar^2}S\right]$$

$$\exp\left(\frac{-x^2\sqrt{mk}}{2\hbar}\right)=-\frac{2mE}{\hbar^2}S\exp\left(\frac{-x^2\sqrt{mk}}{2\hbar}\right)$$

Canceling the first and fifth terms in brackets, dividing both sides by $\exp(x^2(mk)^{1/2} / (2*h\text{-bar}))$, and rearranging:

$$\frac{d^2S}{dx^2}-\frac{2x\sqrt{mk}}{\hbar}\frac{dS}{dx}+\left[\frac{2mE}{\hbar^2}-\frac{\sqrt{mk}}{\hbar}\right]S=0$$

To determine the values of E, one treats S as a power series in order to derive a recursion relation:

$$S=\sum_{n=0}^{\infty}c_nx^n$$

Implementing this:

$$\sum_{n=2}^{\infty}n(n-1)c_nx^{n-2}-\frac{2\sqrt{mk}}{\hbar}\sum_{n=0}^{\infty}nc_nx^n$$

$$+\left[\frac{2mE}{\hbar^2}-\frac{\sqrt{mk}}{\hbar}\right]\sum_{n=0}^{\infty}c_nx^n=0$$

Re-indexing the first term on the left and rewritting as one summation:

$$\sum_{n=0}^{\infty}\left[(n+2)(n+1)c_{n+2}+\left(\frac{2mE}{\hbar^2}-\frac{\sqrt{mk}}{\hbar}-\frac{2n\sqrt{mk}}{\hbar}\right)c_n\right]x^n=0$$

In order for this to be true, the quantity in brackets must be zero for every value of x:

$$(n+2)(n+1)c_{n+2}+\left(\frac{2mE}{\hbar^2}-\frac{\sqrt{mk}}{\hbar}-\frac{2n\sqrt{mk}}{\hbar}\right)c_n=0$$

Rearranging:

$$(n+2)(n+1)c_{n+2} = \left(-\frac{2mE}{\hbar^2} + \frac{\sqrt{mk}}{\hbar} + \frac{2n\sqrt{mk}}{\hbar}\right)c_n$$

In order for the wave function to be finite, one of the three conditions of a wave function, this recursion relation must terminate, otherwise the summation occurs without end. This relation terminates for when n equals a specific value, denoted as v:

$$0 = \frac{2v\sqrt{mk}}{\hbar} + \frac{\sqrt{mk}}{\hbar} - \frac{2mE}{\hbar^2}$$

Rearranging:

$$(2v+1)\sqrt{mk} = \frac{2mE}{\hbar}$$

Isolating energy, produces the sought after eigenvalue expression:

$$E = \hbar\sqrt{\frac{k}{m}}\left(v + \frac{1}{2}\right)$$

This can also be written in terms of the harmonic frequency of the electron's motion, denoted n:

$$E = h\nu\left(v + \frac{1}{2}\right)$$

v is the quantum number for this system; it determines the specific state which the harmonic oscillator electron occupies. Now returning to the function S to derive the wave functions. S is any number of functions belonging to the set known as the Hermite polynomials.

These certain polynomials when multiplied by the asymptotic solution, produce the actual solutions for the wave function. A partial list of the Hermite polynomials is presented below:

upsilon	$S(x)$
0	1
1	$\dfrac{2x^4\sqrt{mk}}{\sqrt{\hbar}}$
2	$\dfrac{4x^2\sqrt[4]{(mk)^2}}{\sqrt{\hbar^2}}-2$
3	$\dfrac{8x^3\sqrt[4]{(mk)^3}}{\sqrt{\hbar^3}}-\dfrac{12x\sqrt[4]{(mk)}}{\sqrt{\hbar}}$
4	$\dfrac{16x^4\sqrt[4]{(mk)^4}}{\sqrt{\hbar^4}}-\dfrac{48x^2\sqrt[4]{(mk)^2}}{\sqrt{\hbar^2}}+12$
5	$\dfrac{32x^5\sqrt[4]{(mk)^5}}{\sqrt{\hbar^5}}-\dfrac{160x^3\sqrt[4]{(mk)^3}}{\sqrt{\hbar^3}}+\dfrac{120x\sqrt[4]{(mk)}}{\sqrt{\hbar}}$

These polynomials are generated by the expression (Mortimer A-45):

$$S_v(x)=(-1)^v \exp\left(\frac{x^2\sqrt{mk}}{\hbar}\right)\frac{\hbar^{v/2}}{(mk)^{v/4}}\frac{d^v}{dx^v}\exp\left(-\frac{x^2\sqrt{mk}}{\hbar}\right)$$

The set of full wave functions for this system is generated by:

$$\psi_v(x)=\left[\frac{\sqrt[4]{mk}}{2^v v!\sqrt{\pi\hbar}}\right]^{1/2} S_v \exp\left(-\frac{x^2\sqrt{mk}}{2\hbar}\right).$$

Graphically demonstrating the wave functions for v = 0,1,2,3,4,5:

Again, just as with the one-dimensional box system, the state, and therefore energy, of the electron is quantized, determined by the quantum number u which is any integer greater than or equal to zero. A surprising result is discovered by analyzing the positions of the wave functions relative to the classical potential energy curve in the above graph. Each wave function extends beyond the potential energy curve, representing the classically allowed maximum displacement of the oscillator. This effect, where the electron occupies space which it is not allowed by classicsal mechanics, is called quantum tunnelling.

This effect does not mean that classical mechanics is necessarily incorrect, rather it helps one to understand that classical mechanics concerns itself with average values pertaining to objects which possess large amounts of energy, relative to electrons.

HYDROGENIC MODEL

The hydrogenic or hydrogen-like atom model is the most complex of the systems described so far. The reason it is called the hydrogen-like atom is because, while it contains only one electron like hydrogen, it may contain any number of nucleons in the atomic nucleus. The electron is present in a spherically symmetrical coulombic field, generated by the positively charged atomic nucleus. It is the common misconception that the nucleus remains stationary while the electron moves about it. In fact, both particles, if one can call them particles anymore, move about the atomic centroid, the center of mass for the atom.

The motion of the electron needs to be studied as relative to this point and not the nucleus. Since it is present in a spherical field, the electron's coordinates relative to the centroid are, surprisingly enough, best shown in spherical coordinates. Also, the effective mass of the electron is changed from its true value, to a value known as the reduced mass, m, which is based on the center of mass coordinates of the electron.

$$\mu = \frac{m_e m_N}{m_e + m_N} \approx 9.10443127198 \times 10^{-31}$$

To begin solving for wave functions and eigenvalues, one starts with the Schrodinger Equation.

$$-\frac{\hbar^2}{2m}\nabla^2\psi + V\psi = E\psi$$

The wave function must now be described in terms of the spherical coordinate variables: r, the distance of the electron from the centroid, ϕ, the angle the radius vector makes with the x-axis when projected onto the xy plane, and θ, the angle the radius vector makes with the z-axis.

Rewriting equation explicitly in spherical coordinates:

$$-\frac{\hbar^2}{2\mu}\left[\frac{1}{r^2}\frac{d}{dr}r^2\frac{d}{dr}+\frac{1}{r^2\sin\theta}\frac{d}{d\theta}+\sin\theta\frac{d}{d\theta}\right.$$

$$\left.+\frac{1}{r^2\sin^2\theta}\frac{d^2}{d\phi^2}\right]\psi-\frac{Ze^2}{4\pi\varepsilon_0^r}\psi=E\psi$$

Notice the potential energy term on the left side which is dependent on 1/r. The Laplacian operator in spherical coordinates is difficult to derive, practically no QM texts show its explicit derivation. To begin solving this monstrous differential equation, the wave function y will need to be broken down into smaller functions. In the first of two reductions, y is broken down into its radial part and its angular part.

$$\psi(r,\phi,\theta)=R(r)Y(\phi,\theta)$$

Implementing equation:

$$-\frac{\hbar^2}{2\mu}\left[\frac{Y}{r^2}\frac{d}{dr}r^2\frac{dR}{dr}+\frac{R}{r^2\sin\theta}\frac{d}{d\theta}\sin\theta\frac{dY}{d\theta}\right.$$

$$\left.+\frac{R}{r^2\sin^2\theta}\frac{d^2Y}{d\phi^2}\right]-\frac{Ze^2}{4\pi\varepsilon_0 r}RY=ERY$$

Notice in each term on the left, only one of the two functions gets differentiated. Dividing by RY and collecting the r^2's and the energy terms:

$$-\frac{\hbar^2}{2\mu r^2}\left[\frac{1}{R}\frac{d}{dr}r^2\frac{dR}{dr}+\frac{1}{Y\sin\theta}\frac{d}{d\theta}\right.$$

$$\sin\theta\frac{dY}{d\theta}+\frac{1}{Y\sin^2\theta}\frac{d^2Y}{d\phi^2}\Bigg]-\frac{Ze^2}{4\pi\varepsilon_0 r}=E$$

Multiplying by $-2\mu r^2$ / (h–bar)^2 and rearranging:

$$\frac{1}{R}\frac{d}{dr}r^2\frac{dR}{dr}+\frac{2\mu r^2}{\hbar^2}\left[E+\frac{Ze^2}{4\pi\varepsilon_0^r}\right]=\eta$$

$$=\frac{-1}{Y\sin\theta}\frac{d}{d\theta}\sin\theta\frac{dY}{d\theta}-\frac{1}{Y\sin^2\theta}\frac{d^2Y}{d\phi^2}$$

In order for this to be true, both sides of this equation must equal each other for all values of r, ϕ, and q. Therefore, each side can be set equal to a constant. This is the exact same procedure used when deriving the Schrodinger Equation. Concentrating on the angular factors first:

$$\eta=\frac{-1}{Y}\left[\frac{1}{\sin\theta}\frac{d}{d\theta}\sin\theta\frac{dY}{d\theta}+\frac{1}{\sin^2\theta}\frac{d^2Y}{d\theta^2}\right]$$

Rearranging:

$$\frac{1}{\sin\theta}\frac{d}{d\theta}\sin\theta\frac{dY}{d\theta}+\frac{1}{\sin^2\theta}\frac{d^2Y}{d\theta^2}+\eta Y=0$$

The angular function must be broken down into two functions, one dependent on ϕ, and the other on θ:

$$Y(\phi,\theta)=\Phi(\phi)\Theta(\theta)$$

Implementing equation:

$$\frac{\Phi}{\sin\theta}\frac{d}{d\theta}\sin\theta\frac{d\Theta}{d\theta}+\frac{\Theta}{\sin^2\theta}\frac{d^2\Phi}{d\theta^2}+\eta\Phi\Theta=0$$

Dividing by $\Phi\Theta$, multiplying by $\sin^2\theta$, separating the factors dependent on f from those dependent on θ, and implementing an arbitrary constant, m^2:

$$\frac{\sin\theta}{\Theta}\frac{d}{d\theta}\sin\theta\frac{d\Theta}{d\theta}+\eta\sin^2\theta=m^2=-\frac{1}{\Phi}\frac{d^2\Phi}{d\phi^2}$$

Concentrating on Φ first produces:

$$\frac{d^2\Phi}{d\phi^2} = -m^2\Phi$$

This expression is similar to the one which appeared when deriving the Schrodinger Equation. The general solutions to Φ are:

$$\Phi(\phi) = Ne^{im\phi}$$

To find the value of N, we normalize:

$$1 = N^2 \int_0^{2\pi} e^{-im\phi} e^{im\phi} d\phi = 2\pi N^2 \Rightarrow N = \frac{1}{\sqrt{2\pi}}$$

To determine the nature of m, we look at the boundary conditions of the function. Since it determines values as a function of the angle, ϕ, we know that it is a periodic function:

$$\frac{1}{\sqrt{2\pi}} e^{im\phi} = \frac{1}{\sqrt{2\pi}} e^{im(\phi+2\pi)} = \frac{1}{\sqrt{2\pi}} e^{im\phi} \left(e^{i\pi}\right)^{2m} \overset{e^{i\pi}=-1}{\Rightarrow} m \in Z$$

The constant m, is a quantum number, meaning it is influential in determining the state of the hydrogen atom system. When the function Φ is operated on by the L_z operator, the z-component of the atom's orbital angular momentum is found.

$$L_z\Phi = \frac{\hbar}{i}\frac{d}{d\phi} e^{im\phi} = m\hbar = \ldots -2\hbar, -\hbar, 0, \hbar, 2\hbar \ldots$$

The angular momentum is a vector, it is best for one to think of an arrow with its base attached at the atomic centroid. The quantum number m determines how far the angular momentum vector goes in the z-direction. If it were not for this, the orientation of the vector would be unknown.

Continuing on to find the other two functions, R and Θ, and their corresponding eigenvalues.....

Returning to equation:

$$\frac{\sin\theta}{\Theta}\frac{d}{d\theta}\sin\theta\frac{d\Theta}{d\theta}+\eta\sin^2\theta=m^2=-\frac{1}{\Phi}\frac{d^2\Phi}{d\phi^2}$$

One can neglect the right-hand side since it has been dealt with already.

$$\frac{\sin\theta}{\Theta}\frac{d}{d\theta}\sin\theta\frac{d\Theta}{d\theta}+\eta\sin^2\theta=m^2$$

Multiplying by Θ, dividing by $\sin^2\theta$, and setting m=0:

$$\frac{1}{\sin\theta}\frac{d}{d\theta}\sin\theta\frac{d\Theta}{d\theta}+\eta\Theta=0$$

As it was stated earlier in this site, many of the complicated differential equations encountered in quantum chemistry are solved by making them resemble equations previously solved. In the case of the angular function Θ, equation resembles the differential equation solved by Adrien-Marie Legendre. To transform it into a workable form requires a few substitutions:

$$y=\cos\theta \qquad \theta=\cos^{-1}y$$

$$dy=-\sin\theta d\theta \quad d\theta=\frac{-dy}{\sqrt{1-y^2}}$$

Utilizing our new substitutions:

$$\frac{d}{dy}\left[\frac{d\Theta}{d\theta}-y^2\frac{d\Theta}{d\theta}\right]+\eta\Theta=0$$

Expanding:

$$\frac{d^2\Theta}{dy^2}-2y\frac{d\Theta}{d\theta}-y^2\frac{d^2\Theta}{d\theta^2}+\eta\Theta=0$$

To solve for the value of η, one can solve this equation by treating Θ as a power series:

$$\Theta(y)=\sum_{n=0}^{\infty}c_n y^n$$

Implementing this:

$$\sum_{n=2}^{\infty} n(n-1)c_n y^{n-2} - 2\sum_{n=0}^{\infty} nc_n y^n - \sum_{n=0}^{\infty} n(n-1)\, c_n y^n + \eta \sum_{n=0}^{\infty} c_n y^n = 0$$

Re-indexing the first term and rewriting the expression as one summation:

$$\sum_{n=0}^{\infty} \left[c_{n+2}\left((N+2)(n+1)\right) - c_n\left(2n + n(n-1) - \eta\right)\right] y^n = 0$$

In order for the above to be true, the quantity in brackets must be zero for all values of y:

$$c_{n+2}\left[(n+2)(n+1)\right] - c_n\left[2n + n(n-1) - \eta\right] = 0$$

Rearranging, one arrives at a recursion relation, similar to the one obtained for the harmonic oscillator:

$$c_{n+2}\left[(n+2)(n+1)\right] = c_n\left[2n + n(n-1) - \eta\right]$$

Also, to simplify:

$$c_{n+2}\left[(n+2)(n+1)\right] = c_n\left[n(n+1) - \eta\right]$$

where $n(n+1) = 2n + n(n-1)$

Now for Θ to be finite, this relation has to terminate. Let the value of n for which it does terminate be labeled l, (there's a reason for using l).

$$0 = l(l+1) - \eta \Rightarrow \eta = l(l+1)$$

The solutions to Θ in the equation presented in equation are (Mortimer):

$$\Theta_{lm}(\theta) = \left[\frac{(2l+l)(l-m)!}{2(l+m)!}\right]^{1/2} \frac{1}{2^l l!} \frac{d^m}{d\theta^m} \frac{1}{\sin^l \theta} \frac{d^l}{d\theta^l} \sin^{2l} \theta$$

The first bracket is a normalizing constant. The reason the letter l was used is because in classical mechanics, l represents angular momentum. The letter l is a quantum number like m, l is member to the set of whole numbers. The quantity l(l+1)*(h-bar)^2 equals the square of the total orbital angular momentum. It is derived from the L^2 operator:

$$\hat{L}^2\Phi\Theta = -\hbar^2\left[\frac{1}{\sin\theta}\frac{d}{d\theta}\sin\theta\frac{d}{d\theta}+\frac{1}{\sin^2\theta}\frac{d^2}{d\phi^2}\right]\Phi\Theta$$

$$= l(l+1)\hbar^2$$

The square-root of l(l+1)*(h-bar)^2 produces the orbital angular momentum for the atom.

Now going way back to equation, one continues to find the remaining function R. Having the value of η:

$$\frac{1}{R}\frac{d}{dr}r^2\frac{dR}{dr}+\frac{2\mu r^2}{\hbar^2}\left[E+\frac{Ze^2}{4\pi\varepsilon_0 r}\right]=l(l+1)$$

Rearranging, multiplying by R, and differentiating:

$$2r\frac{dR}{dr}+r^2\frac{d^2R}{dr^2}+\frac{2\mu r^2}{\hbar^2}\left[E+\frac{Ze^2}{4\pi\varepsilon_0 r}\right]R-l(l+1)R=0$$

Dividing by r^2 and multiplying out the quantity in brackets:

$$\frac{2}{r}\frac{dR}{dr}+\frac{d^2R}{dr^2}+\frac{2\mu E}{\hbar^2}R+\frac{2\mu Ze^2}{4\pi\varepsilon_0 r\hbar}R-\frac{l(l+1)}{r^2}R=0$$

To keep this from becoming too difficult to manipulate, substitutions are made:

$$\alpha^2=-\frac{2\mu E}{\hbar^2}\qquad \beta=\frac{\mu Ze^2}{4\pi\varepsilon_0\alpha\hbar^2}\qquad \rho=2\alpha r$$

Implementing these:

$$\frac{4\alpha}{\rho}\frac{dR}{dr}+\frac{d^2R}{dr^2}-\alpha^2R+\frac{4\alpha^2\beta}{\rho}R-\frac{4\alpha^2l(l+1)}{\rho^2}R=0$$

Since everything is in terms of ρ, one will want to convert the function R from being dependent on r to ρ. The transformations needed to change the previous expression:

$$R(r) \mapsto R(\rho) \frac{dR}{dr} = \frac{dR}{d\rho}\frac{d\rho}{dr} \frac{d^2R}{dr^2} = \frac{d^2R}{d\rho^2}\frac{d\rho^2}{dr^2}$$

Implementing these:

$$\frac{8\alpha^2}{\rho}\frac{dR}{d\rho} + 4\alpha^2 \frac{d^2R}{d\rho^2} - \alpha^2 R + \frac{4\alpha^2\beta}{\rho} R - \frac{4\alpha^2 l(l+1)}{\rho^2} R = 0$$

Dividing by 4* α^2:

$$\frac{2}{\rho}\frac{dR}{d\rho} + \frac{d^2R}{d\rho^2} - \frac{R}{4} + \frac{\beta}{\rho} R - \frac{l(l+1)}{\rho^2} R = 0$$

To ascertain the form of R, the asymptotic solution is taken, similar to the procedure used in the harmonic oscillator situation. The limit of the expression is taken as ρ approaches infinity:

$$\frac{d^2 R_\infty}{d\rho^2} - \frac{R_\infty}{4} = 0$$

The general form of the asymptotic R is:

$$R_\infty = Ne^{-\rho/2}$$

One can state that the actual function R is the product of the asymptotic solution and another function which will be called F here.

$$R = R_\infty F = e^{-\rho/2} F$$

R will now be rewritten as $R_{Inf}F$. Some expressions which will make this possible:

$$\frac{dR}{d\rho} = -\frac{1}{2} e^{-\rho/2} F + e^{-\rho/2} \frac{dF}{d\rho} = e^{-\rho/2}\left[\frac{dF}{d\rho} - \frac{F}{2}\right]$$

Also:

$$\frac{d^2R}{d\rho^2} = \frac{1}{4}e^{-\rho/2}F - e^{-\rho/2}\frac{d^2F}{d\rho^2} = e^{-\rho/2}\left[\frac{F}{4} - \frac{dF}{d\rho} + \frac{d^2F}{d\rho^2}\right]$$

Implementing these:

$$\frac{2}{\rho}\left[\frac{dF}{d\rho} - \frac{F}{2}\right]e^{-\rho/2} + \left[\frac{F}{4} - \frac{dF}{d\rho} + \frac{d^2F}{d\rho^2}\right]e^{-p/2}$$
$$-\frac{F}{4}e^{-\rho/2} + \frac{\beta}{\rho}Fe^{-\rho/2} - \frac{l(l+1)}{\rho^2}Fe^{-\rho/2} = 0$$

The exponential factors can be eliminated, since they never produce a value which would cause the overall expression to equal zero. Simplifing:

$$\frac{2}{\rho}\frac{dF}{d\rho} - \frac{F}{\rho} - \frac{dF}{d\rho} + \frac{d^2F}{d\rho^2} + \frac{\beta}{\rho}F - \frac{l(l+1)}{\rho^2}F = 0$$

Collecting terms of F:

$$\frac{d^2F}{d\rho^2} + \frac{dF}{d\rho}\left[\frac{2}{\rho} - 1\right] + F\left[\frac{\beta}{\rho} - \frac{1}{\rho} - \frac{l(l+1)}{\rho^2}\right] = 0$$

To ascertain the form of F, so one can figure out what R is, it must be broken down. One assumes F is:

$$F = \rho^{\chi}L$$

The exponent on r, χ, is to be determined, L is an unknown function at this point. Some expressions which will be used:

$$\frac{dF}{d\rho} = \chi\rho^{\chi-1}L + \rho^{\chi}\frac{dL}{d\rho}$$

Also:

$$\frac{d^2F}{d\rho^2} = \chi(\chi-1)\rho^{\chi-2}L + 2\chi\rho^{\chi-1}\frac{dL}{d\rho} + \rho^{\chi}\frac{d^2L}{d\rho^2}$$

Implementing these:

$$\left[\chi(\chi-1)\rho^{x-2}L+2\chi\rho^{x-1}\frac{dL}{d\rho}+\rho^{x}\frac{d^2L}{d\rho^2}\right]$$

$$+\left[\chi\rho^{x-1}L+\rho^{x}\frac{dL}{d\rho}\right]\left[\frac{2}{\rho}-1\right]+\left[\rho^{x}L\right]\left[\frac{\beta}{\rho}-\frac{1}{\rho}-\frac{l(l+1)}{\rho^2}\right]=0$$

Expanding and then collecting terms of L:

$$\rho^{x}\frac{d^2L}{d\rho^2}+\left[2\rho^{x-1}-\rho^{x}+2\chi\rho^{x-1}\right]\frac{dL}{d\rho}+$$

$$\left[2\chi\rho^{x-2}-\chi\rho^{x-1}+\beta\rho^{x-1}-\rho^{x-1}-l(l+1)\rho^{x-2}\right.$$

$$\left.+\chi(\chi-1)\rho^{x-2}\right]L=0$$

Dividing by $r^{\chi-2}$ and sorting the χ's in the last bracket by knowing A(A – 1) + 2A = A(A + 1):

$$\rho^{2}\frac{d^2L}{d\rho^2}+\left[2\rho-\rho^{2}+2\chi\rho\right]\frac{dL}{d\rho}+$$

$$\left[\chi(\chi+1)-\chi\rho+\beta\rho-\rho-l(l+1)\right]L=0$$

Grouping the terms with ρ:

$$\rho^{2}\frac{d^2L}{d\rho^2}+\left[(2-\rho+2\chi)\rho\right]\frac{dL}{d\rho}+$$

$$\left[\chi(\chi+1)-l(l+1)+(\beta-\chi-1)\rho\right]L=0$$

To determine the value of χ, the limit is taken as r approaches zero:

$$\left[\chi(\chi+1)-l(l+1)\right]L=0$$

Simplifing:

$$\chi(\chi+1)=l(l+1)\Rightarrow\chi=l$$

This is beneficial. Now χ can be replaced with l:

$$\rho\frac{d^2L}{d\rho^2}+[2-\rho+2l]\frac{dL}{d\rho}+[\beta-l-1]L=0$$

This expression appears similar to the one solved by Edmund Laguerre, a French mathematician:

$$\rho\frac{d^2L_k^j}{d\rho^2}+[j+1-\rho]\frac{dL_k^j}{d\rho}+[k-j]L_k^j=0$$

Making equation conform to Laguerre's equation:

$$\rho\frac{d^2L}{d\rho^2}+[2(l+1)-\rho]\frac{dL}{d\rho}+[\beta-(l+1)L]=0$$

Looking at the first brackets in each equation, one can draw similarities between the two:

$$j+1-\rho=2(l+1)-\rho\Rightarrow j=2l+1$$

Also, looking at the second brackets:

$$k-j=\beta-l-1\Rightarrow\beta=k-l$$

Laguerre found the solutions to L correspond to the following expression:

$$L_k^j=\frac{d^j}{d\rho^j}\left[e^{\rho}\frac{d^k}{d\rho^k}\left[\rho^k e^{-\rho}\right]\right]j,k\in N$$

One will notice by looking at this expression, that j and k must be natural numbers. Also, for L to be nonzero:

$$j=2l+1\leq k\Rightarrow l\leq k$$

The fact that l is less than or equal to k is important for realizing that beta in equation is equal to a positive integer. This integer can be represented as n, the quantum number used in the particle-in-a-box system. Symbolizing this:

$$\beta=n=\frac{\mu Ze^2}{4\pi\varepsilon_0\alpha\hbar^2}$$

Using this, the full radial function R, can finally be constructed:

$$R_{nl}(\rho)=\left[\left(\frac{2Z}{na_0}\right)^3\frac{(n-l-1)!}{2n[(n+l)!]^3}\right]^{1/2} e^{-\rho/2}\rho^l\frac{d^{2l+1}}{d\rho^{2l+1}}\left[e^{\rho}\frac{d^{n+1}}{d\rho^{n+1}}\left(\rho^{n+1}e^{-\rho}\right)\right]$$

The first bracket is a normalizing constant.

Rearranging equation so α is isolated, it can be compared with the value of α^2 in equation:

$$\alpha=\frac{\mu Ze^2}{4\pi\varepsilon_0 n\hbar^2}\Rightarrow-\frac{2\mu E}{\hbar^2}=\alpha^2=\frac{\mu^2Z^2e^4}{16\pi^2\varepsilon_0^2n^2\hbar^4}$$

Rearranging for energy produces the sought after energy expression which determines the energy states of the hydrogen-like atom system:

$$E=\frac{\mu Z^2e^4}{2(4\pi\varepsilon_0)^2\,n^2\hbar^2}=-2.1787\times10^{-18}J\frac{Z^2}{n^2}$$

This is dependent on Z, the number of charged nucleons in the atom, and n, the principal quantum number which is any integer greater than or equal to one.

Just as with the first two systems, the hydrogen-like atom possesses quantized energy states. An electron in this situation, will only make certain energy transitions, thereby emitting or absorbing only discrete amounts of energy. The overall wave function for the electron is dependent on three quantum numbers; n, l, and m. A generalized expression of the wave function can be written:

$$\psi(r,\phi,\theta)=R_{nl}(r)\Phi_m(\phi)\Theta_{lm}(\theta)$$

Wave functions of single electrons are called orbitals, indirectly making reference to Niels Bohr's model of orbiting electrons. An orbital region is the region of space around the atomic nucleus which an electron is most likely to occupy. The principal quantum number n indicates the energy level an electron is in. It is any integer greater than or equal to one. The azimuthal quantum number l indicates the orbital angular momentum of the electron and therefore the shape of the orbital region. The magnetic quantum number m indicates the z-component of the orbital angular momentum and therefore indicates the orientation of the orbital region.

As discovered by P.A.M. Dirac, electrons possess a quantum analogue to intrinsic angular momentum, called spin. This fact is derived from relativistic quantum mechanics, not from the Schrodinger Equation. A fourth quantum number m_s indicates whether this intrinsic angular momentum is +h-bar/2 or -h-bar/2. Two electrons can occupy the same orbital or in other words have the same wave function, pending their intrinsic angular momentum projection vectors oppose each other.

The hydrogen-like atom model cannot be directly applied to polyelectronic systems. In such systems, the electrons exert repulsive forces against each other, greatly effecting the energy and orbitals of the system. Various approximation methods can be implemented to determine the states of such systems.

HYDROGENIC ISOSURFACES

The following are isosurface representations for hydrogenic wavefunctions. An orbital isosurface is defined by the points where the wavefunction is equal to an arbitrary constant. Although in reality, since the true hydrogenic orbitals are scalar fields of complex numbers, there is no one good way to represent them. The ones presented here are formed from linear combinations which represent the real component of each orbital.They demonstrate the general shapes of the orbital regions, as determined by the angular momentum quantum number, l, and the angular momentum projection quantum number, m_l. The colours indicate the relative phase of the orbital, where the wavefunction is positive and negative.

THE HELIUM ATOM

The helium atom is an example of a polyelectronic system. It is comprised of a nucleus with a atomic number of Z=2 and two electrons. This system cannot be treated in the same manner that the hydrogenic model was due to the interactions among the two electrons. One begins by writing the hamiltonian for the atom.

$$H^{He} = H^1 + H^2 + \frac{e^2}{4\pi\varepsilon_0 r_{12}}$$

The total hamiltonian is simply comprised of the one electron hamiltonains devised earlier and also an operator to account for electron-electron repulsion, called the coulomb operator.

An eigenfunction which will commute with the helium hamiltonian constructed is a product of two one-electron atomic orbitals, one for each of the electrons in ^{2}He.

$$\psi_1^{He} = \psi_1(1)\phi_2(2)\psi_2^{He} = \phi_1(2)\phi_2(1)$$

Two possible wavefunctions are possible, depending on which electron occupies which one-electron orbital. Caution is warrented here, these wavefunctions are obviously not the exact solutions due to electron-electron interactions. However in the limiting case of non-interaction, these are correct and they provide a good starting point anyway. As will be shown, for states other than the ground state, the two one-electron orbitals will differ causing the atom to be subject to additional electronic phenomena. The one-electron hamiltonians are of the form:

$$H^i = \frac{-\hbar^2}{2m}\nabla^2 - \frac{2e^2}{4\pi\varepsilon_0 r_i}$$

These consist of the kinetic energy operator and the nuclear electrostatic potnetial operator. Since there exist two possible wavefunctions, all possible combinations of integrals need to be accounted for when operating with the hamiltonian.

$$H_{\mu\nu}^{He} = \left\langle \psi_{\mu}^{He} \middle| H^{He} \middle| \psi_{\nu}^{He} \right\rangle$$

An easy way of keeping track of the integrals is to treat the hamiltonian as a matrix.

$$H_{\mu\nu}^{He} = \begin{pmatrix} H_{11} & H_{12} \\ H_{21} & H_{22} \end{pmatrix}$$

The elements of this matrix corrsespond to operations of various combinations of the possible wavefunctions. We assume the two helium wavefunctions are orthogonal since they are centered on the same nucleus, meaning the overlap matrix is of the form:

$$S_{\mu\nu}^{He} = \delta_{\mu\nu}$$

Continuing, one turns to the Schrodinger Equation, which is of the form:

$$\sum_{\nu} H_{\mu\nu}^{He} \psi_{\nu}^{He} = E \sum_{\nu} S_{\mu\nu}^{He} \psi_{\nu}^{He}$$

This can be rearranged into the familiar eigenvalue equation:

$$\det \left| H_{\mu\nu}^{He} - E S_{\mu\nu}^{He} \right| = 0$$

Unique solutions exist for when the determinant is zero. Knowing the overlap matrix, all one needs to determine is the form of the hamiltonian matrix. Each of the matrix elements needs to be solved in turn. For element H_{11}, one has:

$$H_{11} = \left\langle \phi_1(1)\phi_2(2) \middle| H^1 + H^2 + \frac{e^2}{4\pi\varepsilon_0 r_{12}} \middle| \phi_1(1)\phi_2(2) \right\rangle$$

Each of the one-electron atomic orbitals will commute with its respective one-electron hamiltonian and the coulomb operator. As a result of the integral, one obtains the following.

$$H_{11} = E^1 + E^2 + \frac{e^2}{4\pi\varepsilon_0} \left\langle \phi_1(1)\phi_2(2) \middle| \frac{1}{r_{12}} \middle| \phi_1(1)\phi_2(2) \right\rangle$$

One sees the result is just the energy of each electron with reference to the bare nuclear field plus a third term which is the coulomb term. The coulomb term is always denoted as J in literature, which will also be done here.

$$J = \frac{e^2}{4\pi\varepsilon_0}\left\langle \phi_1(1)\phi_2(2)\left|\frac{1}{r_{12}}\right|\phi_1(1)\phi_2(2)\right\rangle$$

The solution to the coulomb integral is difficult to derive and can only be solved for through computational methods covered later in this site. For now it is merely important to recognize its physical interpretation, that being the interelectron repulsion. As one can prove independently, the H_{22} term is equal to the H_{11} term.

$$H_{22} = H_{11}$$

Next, one needs to consider the other two terms of the hamiltonian matrix, H_{21} and H_{12}. Setting up the integral for H_{21} results in:

$$H_{21} = \left\langle \phi_1(2)\phi_2(1)\left|H^1 + H^2 + \frac{e^2}{4\pi\varepsilon_0 r_{12}}\right|\phi_1(1)\phi_2(2)\right\rangle$$

Solving results in a solution which looks similar to the ones derived for the first two terms solved for, but with very important differences.

$$H_{21} = E^1\left\langle \psi_2^{He} \mid \psi_1^{He}\right\rangle + E^2\left\langle \psi_2^{He} \mid \psi_1^{He}\right\rangle + \frac{e^2}{4\pi\varepsilon_0}\left\langle \psi_2^{He}\left|\frac{1}{r_{12}}\right|\psi_1^{He}\right\rangle$$

Since the orthogonality restriction has been placed on this system, one immediately recognizes the first two terms are zero. The remaining term is called the exchange integral. It represents the ability of two electrons with parallel spins to exchange places. In the ground state of helium, the two electrons occupy the same space orbital and therefore have opposing spins. This means the two cannot exchange places.

However, in the case of excited states of helium, the electrons will occupy different space orbitals and therefore have parallel spins. This means the two can in fact exchange places with each other. This term is always denoted K in literature.

$$K = \frac{e^2}{4\pi\varepsilon_0}\left\langle \psi_2^{He} \left| \frac{1}{r_{12}} \right| \psi_1^{He} \right\rangle$$

Again, one can prove independently that the matrix elements H_{21} and H_{12} are equal.

$$H_{12} = H_{21}$$

Now one has everything to solve for the energy of the atom. Filling in the eigenvalue equation with the terms one has derived, the result is:

$$\det\begin{vmatrix} E^1 + E^2 + J - E & K \\ K & E^1 + E^2 + J - E \end{vmatrix} = 0$$

Taking the determinant leaves one with a simple polynomial equation.

$$\left(E^1 + E^2 + J - E\right)^2 - K^2 = 0$$

Rearranging this leaves one with the final expression of the energy for the helium atom.

$$E = E^1 + E^2 + J \pm K$$

There exist two solutions, however only the one containing the minus sign is physically important, since the universe has a tendency to keep energy at a minimum. The energy of the helium atom is therefore the sum of the two electron energies as influenced only by the nucleus, the energy of the interelectron repulsion, and the stabalization energy dervied from electron exchange.

Chapter 2

Atoms and Atomic Spectroscopy

FREQUENCY, WAVENUMBER, WAVELENGTH

The photon energy due to an electron transition between an upper atomic level k (of energy E_k) and a lower level i is

$$\Delta E = E_k - E_i = h\nu = hc\sigma = hc/\lambda_{vac}$$

where ν is the frequency, σ the wavenumber in vacuum, and λ_{vac} the wavelength in vacuum. The most accurate spectroscopic measurements are determinations of transition frequencies, the unit being the Hertz (1 Hz = cycle/sec) or one of its multiples. A measurement of any one of the entities frequency, wavenumber, or wavelength (in vacuum) is an equally accurate determination of the others since the speed of light is exactly defined. The most common wavelength units are the nanometer (nm), the Ångström (1 = 10^{-1} nm) and the micrometer (μm). The SI wavenumber unit is the inverse meter, but in practice wavenumbers are usually expressed in inverse centimeters: 1 cm^{-1} = 10^2 m^{-1}, equivalent to 2.997 924 58 × 10^4 MHz.

In addition to frequency and wavenumber units, atomic energies are often expressed in electron volts (eV). One eV is the energy associated with each of the following quantities:

2.417 989 40(21) × 10^{14} Hz

8 065.544 45(69) cm^{-1}

1 239.841 91(11) nm

11 604.505(20) K (kelvin)

1.602 176 53(14) × 10^{-19} J (joule)

We note that the basic unit of temperature, the kelvin, is equivalent to about 0.7 cm^{-1}, i.e., the value of the Boltzmann constant k expressed in wavenumber units per kelvin is 0.695 035 6(12) cm^{-1}/K. One reason for citing this particular equivalency involving the internationally accepted symbol for the kelvin (K) is to suggest that use of the letter K as a symbol for 1 cm^{-1} (1 kayser) should be discontinued.

The unit of energy in the system of atomic units (a.u.) often used for theoretical calculations is the hartree, which is equal to 2 rydbergs. The rydberg for an atom having nuclear mass M is

$$1\ \text{Ry} = R_M = M(M + m_e)^{-1}\ R_\infty \quad ,$$

with

$$R_\infty = m_e c\alpha^2/(2h) = 10\ 973\ 731.568\ 525(73)\ \text{m}^{-1} \quad ,$$

which is equivalent to 13.605 692 3(12) eV. The Rydberg constant R_∞ is thus the limit value for infinite nuclear mass.

ATOMIC STATES, SHELLS, AND CONFIGURATIONS

A one-electron atomic *state* is defined by the quantum numbers nlm_lm_s or $nljm_j$, with n and l representing the principal quantum number and the orbital angular momentum quantum number, respectively. The allowed values of n are the positive integers, and $l = 0, 1, ..., n-1$. The quantum number j represents the angular momentum obtained by coupling the orbital and spin angular momenta of an electron, i.e., $j = l + s$, so that $j = l \pm \frac{1}{2}$. The magnetic quantum numbers m_l, m_s, and m_j represent the projections of the corresponding angular momenta along a particular direction; thus, for example, $m_l =$

$-l, -l+1 \ldots l$ and $m_s = \pm\frac{1}{2}$.

The central field approximation for a many-electron atom leads to wave functions expressed in terms of products of such one-electron states. Those electrons having the same principal quantum number n belong to the *shell* for that number. Electrons having both the same n value and l value belong to a *subshell,* all electrons in a particular subshell being *equivalent.* The notation for a *configuration* of N equivalent electrons is nl^N, the superscript usually being omitted for $N = 1$. A configuration of several subshells is written as $nl^N n' l'^M$ ... . The numerical values of l are replaced by letters in writing a configuration, according to the code *s, p, d* for $l = 0, 1, 2$ and *f, g, h* ... for $l = 3, 4, 5$ ..., the letter j being omitted.

The Pauli exclusion principle prohibits atomic states having two electrons with all four quantum numbers the same. Thus the maximum number of equivalent electrons is $2(2l + 1)$. A subshell having this number of electrons is *full, complete,* or *closed,* and a subshell having a smaller number of electrons is *unfilled, incomplete,* or *open.* The $3p^6$ configuration thus represents a full subshell and $3s^2\ 3p^6\ 3d^{10}$ represents a full shell for $n = 3$.

The *parity* of a configuration is *even* or *odd* according to whether $\Sigma_i l_i$ is even or odd, the sum being taken over all electrons.

HYDROGEN AND HYDROGEN-LIKE IONS

The quantum numbers n, l, and j are appropriate. A particular *level* is denoted either by nl_j or by $nl\ {}^2L_J$ with $L = l$ and $J = j$. The latter notation is somewhat redundant for one-electron spectra, but is useful for consistency with more complex structures. The L values are written with the same letter code used for l values, but with roman capital letters. The *multiplicity* of the *L term* is equal to $2S + 1 = 2s + 1 = 2$. Written as a superscript, this number expresses the *doublet* character of the structure: each term for $L \geq 1$ has two levels, with $J = L \pm \frac{1}{2}$, respectively.

The Coulomb interaction between the nucleus and the single electron is dominant, so that the largest energy separations are associated with levels having different n. The hyperfine splitting of the ^{1}H 1s ground level results from the interaction of the proton and electron magnetic moments and gives rise to the famous 21 cm line.

The separations of the $2n - 1$ excited levels having the same n are largely determined by relativistic contributions, including the spin-orbit interaction, with the result that each of the $n - 1$ pairs of levels having the same j value is almost degenerate; the separation of the two levels in each pair is mainly due to relatively small Lamb shifts.

ALKALIS AND ALKALI-LIKE SPECTRA

In the central field approximation there exists no angular-momentum coupling between a closed subshell and an electron outside the subshell, since the net spin and orbital angular momenta of the subshell are both zero. The *nlj* quantum numbers are, then, again appropriate for a single electron outside closed subshells.

However, the electrostatic interactions of this electron with the core electrons and with the nucleus yield a strong l-dependence of the energy levels. The differing extent of "core penetration"" for *ns* and *np* electrons can in some cases, for example, give an energy difference comparable to or exceeding the difference between the *np* and $(n + 1)p$ levels. The spin-orbit fine-structure separation between the *nl* $(l > 0)$ levels having $j = l - \frac{1}{2}$ and $l + \frac{1}{2}$, respectively, is relatively small.

HELIUM AND HELIUM-LIKE IONS; *LS* COUPLING

The energy structure of the normal 1*snl* configurations is dominated by the electron-nucleus and electron-electron Coulomb contributions. In helium and in helium-like ions of the lighter elements, the separations of levels having the same n and having $l = s$, p, or d are mainly determined by direct and exchange electrostatic interactions between the electrons - the spin-orbit, spin-other orbit, and other relativistic contributions are much smaller. This is the condition for *LS* coupling, in which:

a. The orbital angular momenta of the electrons are coupled to give a total orbital angular momentum $L = \sum_i l_i$.
b. The spins of the electrons are coupled to give a total spin $S = \sum_i s_i$.

The combination of a particular *S* value with a particular *L* value comprises a spectroscopic *term*, the notation for which is ^{2S+1}L. The quantum number $2S + 1$ is the *multiplicity* of the term. The *S* and *L* vectors are coupled to obtain the total angular momentum, $J = S + L$, for a *level* of the term; the level is denoted as $^{2S+1}L_J$.

For 1*snl* configurations, $L = l$ and $S = 0$ or 1, i.e., the terms are *singlets* ($S = 0$) or *triplets* ($S = 1$). As examples of the He I structure, the ionization energy (energy required to remove one of the 1*s* electrons in the $1s^2$ ground configuration) is 24.5874 eV, the 1*s*2*s* $^3S - {}^1S$ separation is 0.7962 eV, the 1*s*2*p* $^3P^o - {}^1P^o$ separation is 0.2539 eV, and the 1*s*2*p* $^3P^o_2 - {}^3P^o_0$ fine-structure spread is only 1.32×10^{-4} eV.

HIERARCHY OF ATOMIC STRUCTURE IN *LS* COUPLING

The centrality of *LS* coupling in the analysis and theoretical interpretation of atomic spectra has led to the acceptance of notations and nomenclature well adapted to discussions of particular structures and spectra. The main elements of the nomenclature are shown in the table below, most of the structural entities having already been defined in the above discussions of simple spectra. The quantum numbers in the table represent a full description for complex configurations, and the accepted names for transitions between the structural elements are also given.

Table: Atomic structural hierarchy in *LS* coupling and names for the groups of all transitions between structural entities.

Structural entity	Quantum numbers	Group of all transitions
Configuration	$(n_i l_i)^{N_i}$	Transition array
Polyad	$(n_i l_i)^{N_i}\, \gamma\, S_1\, L_1\, nl\, S\, L,\, S\, L' \ldots$	Supermultiplet
Term	$(n_i l_i)^{N_i}\, \gamma\, S\, L$	Multiplet
Level	$(n_i l_i)^{N_i}\, \gamma\, S\, L\, J$	Line
State	$(n_i l_i)^{N_i}\, \gamma\, S\, L\, J\, M$	Line component

The configuration may include several open subshells, as indicated by the *i* subscripts. The letter γ represents any additional quantum numbers, such as ancestral terms, necessary to specify a particular term.

As an example, the Ca I $3d4p\,^3D^\circ_2$ *level* belongs to the $^3D^\circ$ *term* which, in turn, belongs to the $3d4p\ ^3(P^\circ\ D^\circ\ F^\circ)$ triplet *triad*. The *3d4p configuration* also has a $^1(P^\circ\ D^\circ\ F^\circ)$ singlet *triad*. The *3d4s* configuration has only *monads*, one ^1D and one ^3D. The $3d4s\ ^3D_2 - 3d4p\ ^3D^\circ_3$ *line* belongs to the corresponding $^3D - {}^3D^\circ$ triplet *multiplet*, and this multiplet belongs to the great Ca I $3d4s\ ^3D - 3d4p\ ^3(P^\circ\ D^\circ\ F^\circ)$ *supermultiplet* of three triplet multiplets discussed by Russell and Saunders in their classic paper on the alkaline-earth spectra. The *3d4s – 3d4p transition array* includes both the singlet and triplet supermultiplets, as well as any (*LS*-forbidden) *intercombination* or *intersystem* lines arising from transitions between levels of the singlet *system* and those of the triplet *system*. The order of the two terms in the transitions as written above, with the lower-energy term on the left, is standard in atomic spectroscopy. Examples of notations for complex configurations are given in Notations for Different Coupling Schemes.

ALLOWED TERMS OF LEVELS FOR EQUIVALENT ELECTRONS

LS Coupling

The allowed *LS* terms of a configuration consisting of two nonequivalent groups of electrons are obtained by coupling the *S* and *L* vectors of the groups in all possible ways, and the procedure may be extended to any number of such groups. Thus the allowed terms for any configuration can be obtained from a table of the allowed terms for groups of equivalent electrons.

The configuration l^N has more than one allowed term of certain *LS* types if $l > 1$ and $2 < N < 4l$ ($d^3 - d^7$, $f^3 - f^{11}$, etc.). The recurring terms of a particular *LS* term type from d^N and f^N configurations are assigned sequential index numbers in the tables of Nielson and Koster; the index numbers stand for additional numbers having group-theoretical significance that serve to differentiate the recurring terms, except for a few terms of f^5 and f^9, f^6 and f^8, and f^7. These remaining terms, which occur only in pairs, are further labeled *A* or *B* to indicate Racah's separation of the two terms.

The index numbers of Nielson and Koster are in practice the most frequently used labels for the recurring terms of f^N configurations. Use of their index numbers for the recurring terms of d^N configurations has perhaps the disadvantage of substituting an arbitrary number for a quantum number (the seniority) that itself distinguishes the recurring terms in all cases. The actual value of the seniority number is rarely needed, however, and a consistent notation for the d^N and f^N configurations is desirable. A table of the allowed *LS* terms of the l^N electrons for $l \leq 3$ is given in Ref., with all recurring terms having the index numbers of Nielson and Koster as a following on-line integer. The theoretical group labels are also listed. Thus the $d^3\ {}^2D$ term having seniority 3 is designated ${}^2D2$, instead of ${}^2_3D$, in this scheme; and the level having $J = {}^3/_2$ is designated ${}^2D_{3/2}2$.

jj Coupling

The allowed *J* values for a group of *N* equivalent electrons having the same *j* value, l_j^N, are given in the table below for $j = {}^1/_2, {}^3/_2, {}^5/_2$, and ${}^7/_2$ (sufficient for $l \leq 3$). The $l^4_{7/2}$ group has two allowed levels for each of the *J* values 2 and 4. The subscripts distinguishing the two levels in each case are the seniority numbers.

Allowed J values for l_j^N equivalent electrons (jj coupling).

ljN	Allowed J values
$l_{1/2}$	1/2
$l^2{}_{1/2}$	0
$l_{3/2}$ and $l^3{}_{3/2}$	3/2
$l^2{}_{3/2}$	0, 2
$l^4{}_{3/2}$	0
$l_{5/2}$ and $l^5{}_{5/2}$	5/2
$l^2{}_{5/2}$ and $l^4{}_{5/2}$	0, 2, 4
$l^3{}_{5/2}$	3/2, 5/2, 9/2
$l^6{}_{5/2}$	0
$l_{7/2}$ and $l^7{}_{7/2}$	7/2
$l^2{}_{7/2}$ and $l^6{}_{7/2}$	0, 2, 4, 6
$l^3{}_{7/2}$ and $l^5{}_{7/2}$	3/2, 5/2, 7/2, 9/2, 11/2, 15/2
$l^4{}_{7/2}$	0, $_22$, $_42$, $_24$, $_44$, 5, 6, 8
$l^8{}_{7/2}$	0

The allowed levels of the configuration nl^N may be obtained by dividing the electrons into sets of two groups $nl^Q_{l+1/2} nl^R_{l-1/2}$, $Q + R = N$. The possible sets run from $Q = N - 2l$ (or zero if $N > 2l$) up to $Q = N$ or $Q = 2l + 2$, whichever is smaller. The (degenerate) levels for a set with both Q and R nonzero have wave functions defined by the quantum numbers (αJ_1, βJ_2) J, with J_1 and J_2 deriving from the Q and R groups, respectively. The symbols α and β represent any additional quantum numbers required to identify levels. The J values of the allowed levels for each (αJ_1, βJ_2) subset are obtained by combining J_1 and J_2 in the usual way.

NOTATIONS FOR DIFFERENT COUPLING SCHEMES

The following examples make clear the meaning of the different coupling-scheme notations. Not all the configurations in the examples have been identified experimentally, and some of the examples of a particular coupling scheme given for heuristic purposes may be physically inappropriate. Cowan describes the physical conditions for the different coupling schemes and gives experimental examples.

LS COUPLING (RUSSELL-SAUNDERS COUPLING)

Some of the examples given below indicate notations bearing on the order of coupling of the electrons.

1. $3d^{7}\ {}^{4}F_{7/2}$
2. $3d^{7}({}^{4}F)4s4p({}^{3}P^{\circ})\ {}^{6}F^{\circ}{}_{9/2}$
3. $4f^{7}({}^{8}S^{\circ})6s6p^{2}({}^{4}P)\ {}^{11}P^{\circ}{}_{5}$
4. $3p^{5}({}^{2}P^{\circ})3d^{2}({}^{1}G)\ {}^{2}F^{\circ}{}_{7/2}$
5. $4f^{10}({}^{3}K2)6s6p({}^{1}P^{\circ})\ {}^{3}L^{\circ}{}_{6}$
6. $4f^{7}({}^{8}S^{\circ})5d\ ({}^{7}D^{\circ})6p\ {}^{8}F^{\circ}{}_{13/2}$
7. $4f^{7}({}^{8}S^{\circ})5d\ ({}^{9}D^{\circ})6s\ ({}^{8}D^{\circ})7s\ {}^{9}D^{\circ}{}_{5}$
8. $4f^{7}({}^{8}S^{\circ})5d\ ({}^{9}D^{\circ})6s6p({}^{3}P^{\circ})\ {}^{11}F_{8}$
9. $4f^{7}({}^{8}S^{\circ})5d^{2}({}^{1}G)\ ({}^{8}G^{\circ})6p\ {}^{7}F_{0}$
10. $4f({}^{2}F^{\circ})\ 5d^{2}({}^{1}G)6s\ ({}^{2}G)\ {}^{1}P^{\circ}{}_{1}$

In the second example, the seven $3d$ electrons couple to give a ${}^{4}F$ term, and the $4s$ and $4p$ electrons couple to form the ${}^{3}P^{\circ}$ term; the final ${}^{6}F^{\circ}$ term is one of nine possible terms obtained by coupling the ${}^{4}F$ grandparent and ${}^{3}P^{\circ}$ parent terms. The next three examples are similar to the second. The meaning of the index number 2 following the ${}^{3}K$ symbol in the fifth example is explained in the section *LS* Coupling.

The coupling in example 6 is appropriate if the interaction of the $5d$ and $4f$ electrons is sufficiently stronger than the $5d$-$6p$ interaction. The ${}^{7}D^{\circ}$ parent term results from coupling the $5d$ electron to the ${}^{8}S^{\circ}$ grandparent, and the $6p$ electron is then coupled to the ${}^{7}D^{\circ}$ parent to form the final ${}^{8}F$ term. A space is inserted between the $5d$ electron and the ${}^{7}D^{\circ}$ parent to emphasize that the latter is formed by coupling a term (${}^{8}S^{\circ}$) listed to the left of the space. Example 7 illustrates a similar coupling order carried to a further stage; the ${}^{8}D^{\circ}$ parent term results from the coupling of the $6s$ electron to the ${}^{9}D^{\circ}$ grandparent.

Example 8 is similar to examples 2 through 5, but in 8 the first of the two terms that couple to form the final ^{11}F term, i.e., the ^{9}D° term, is itself formed by the coupling of the 5*d* electron to the ^{8}S° core term. Example 9 shows an ^{8}G° parent term formed by coupling the ^{8}S° and ^{1}G grandparent terms. A space is again used to emphasize that the following (^{8}G°) term is formed by the coupling of terms listed before the space.

A different order of coupling is indicated in the final example, the $5d^{2}$ ^{1}G term being coupled first to the external 6*s* electron instead of directly to the 4*f* core electron. The 4*f* (^{2}F°) core term is isolated by a space to denote that it is coupled (to the $5d^{2}$(^{1}G)6*s* ^{2}G term) only after the other electrons have been coupled. The notation in this particular case (with a single 4*f* electron) could be simplified by writing the 4*f* electron after the ^{2}G term to which it is coupled. It appears more important, however, to retain the convention of giving the core portion of the configuration first.

The notations in examples 1 through 5 are in the form recommended by Russell, Shenstone, and Turner, and used in both the *Atomic Energy States* and *Atomic Energy Levels*, compilations. The spacings used in the remaining examples allow different orders of coupling of the electrons to be indicated without the use of additional parentheses, brackets, etc.

Some authors assign a short name to each (final) term, so that the configuration can be omitted in tables of classified lines, etc. The most common scheme distinguishes the low terms of a particular *SL* type by the prefixes *a, b, c, ...,* and the high terms by *z, y, x, ...*

JJ COUPLING OF EQUIVALENT ELECTRONS

This scheme is used, for example, in relativistic calculations. The lower-case *j* indicates the angular momentum of one electron ($j = l \pm {}^{1}/_{2}$) or of each electron in an l_j^N group. Various ways of indicating which of the two possible *j* values applies to such a group without writing the *j*-value subscript have been used by different authors; we give the j values explicitly in the examples below. We use the symbols J_i and j to represent total angular momenta.

1. $(6p\,^2{}_{1/2})_0$
2. $(6p\,^2{}_{1/2}\,6p_{3/2})^{\circ}{}_{3/2}$
3. $(6p\,^2{}_{1/2}\,6p\,^2{}_{3/2})_2$
4. $4d\,^3{}_{5/2}\,4d\,^2{}_{3/2}\,(^9/_2, 2)_{11/2}$

The relatively large spin-orbit interaction of the $6p$ electrons produces jj-coupling structures for the $6p\,^2$, $6p\,^3$, and $6p\,^4$ ground configurations of neutral Pb, Bi, and Po, respectively; the notations for the ground levels of these atoms are given as the first three examples above. The configuration in the first example shows the notation for equivalent electrons having the same j value l_j^N, in this case two $6p$ electrons each having $j = {}^1/_2$. A convenient notation for a particular level ($J = 0$) of such a group is also indicated. The second example extends this notation to the case of a $6p\,^3$ configuration divided into two groups according to the two possible j values. A similar notation is shown for the $6p\,^4$ level in the third example; this level might also be designated $(6p\,^{-2}{}_{3/2})_2$, the negative superscript indicating the two $6p$ holes. The $(J_1, J_2)_J$ term and level notation shown on the right in the fourth example is convenient because each of the two electron groups $4d\,^3{}_{5/2}$ and $4d\,^2{}_{3/2}$ has more than one allowed total J_i value. The assumed convention is that J_1 applies to the group on the left ($J_1 = {}^9/_2$ for the $4d\,^3{}_{5/2}$ group) and J_2 to that on the right.

$J_1\ j$ or $J_1 J_2$ Coupling

1. $3d\,^9(^2D_{5/2})4p_{3/2}\ (^5/_2, {}^3/_2)^{\circ}{}_3$
2. $4f^{11}(^2H^{\circ}{}_{9/2}\,2)6s6p(^3P^{\circ}{}_1)\ (^9/_2, 1)_{7/2}$
3. $4f^9(^6H^{\circ})5d\ (^7H^{\circ}{}_8)6s6p(^3P^{\circ}{}_0)\ (8,0)_8$
4. $4f^{12}(^3H_6)\,5d(^2D)6s6p(^3P^{\circ})\ (^4F^{\circ}{}_{3/2})\ (6, {}^3/_2)^{\circ}{}_{13/2}$
5. $5f^4(^5I_4)6d_{3/2}\ (4,{}^3/_2)_{11/2}7s7p(^1P^{\circ}{}_1)\ (^{11}/_2, 1)^{\circ}{}_{9/2}$
6. $5f^4{}_{7/2}5f^5{}_{5/2}\,(8,{}^5/_2)^{\circ}{}_{21/2}7p_{3/2}\ (^{21}/_2, {}^3/_2)_{10}$
7. $5f^3{}_{7/2}\,5f^3{}_{5/2}\,(^9/_2, {}^9/_2)_9 7s7p(^3P^{\circ}{}_2)\ (9,2)^{\circ}{}_7$

The first five examples all have core electrons in *LS* coupling, whereas *jj* coupling is indicated for the 5*f* core electrons in the last two examples. Since the J_1 and J_2 values in the final (J_1, J_2) term have already been given as subscripts in the configuration, the (J_1, J_2) term notations are redundant in all these examples. Unless separation of the configuration and final term designations is desired, as in some data tables, one may obtain a more concise notation by simply enclosing the entire configuration in brackets and adding the final *J* value as a subscript.

Thus, the level in the first example can be designated as $[3d^9(^2D_{5/2})\,4p_{3/2}]^\circ_3$. If the configuration and coupling order are assumed to be known, still shorter designations may be used; for example, the fourth level above might then be given as $[(^3H_6)\,(^3P^\circ)\,(^4F^\circ_{3/2})]_{13/2}$ or $(^3H_6,\,^3P^\circ,\,^4F^\circ_{3/2})_{13/2}$. Similar economies of notation are of course possible, and often useful, in all coupling schemes.

J_1l or J_1L_2 Coupling (J_1K Coupling)

1. $3p^5(^2P^\circ_{1/2})5g\ ^2[^9/_2]^\circ_5$
2. $4f^2(^3H_4)5g\ ^2[3]_{5/2}$
3. $4f^{13}(^2F^\circ_{7/2})5d^2(^1D)\ ^1[^7/_2]^\circ_{7/2}$
4. $4f^{13}(^2F^\circ_{5/2})5d6s(^3D)\ ^3[^9/_2]^\circ_{11/2}$

The final terms in the first two examples result from coupling a parent-level J_1 to the orbital angular momentum of a 5*g* electron to obtain a resultant *K*, the *K* value being enclosed in brackets. The spin of the external electron is then coupled with the *K* angular momentum to obtain a pair of *J* values, $J = K \pm {}^1/_2$ (for *K* '" 0).

The multiplicity (2) of such pair terms is usually omitted from the term symbol, but other multiplicities occur in the more general J_1L_2 coupling (examples 3 and 4). The last two examples are straightforward extensions of J_1l coupling, with the L_2 and S_2 momenta of the "external"' term (^1D and ^3D in examples 3 and 4, respectively) replacing the *l* and *s* momenta of a single external electron.

LS_1 Coupling (LK Coupling)

1. $3s\,^2 3p(^2P^\circ)4f$ G $^2[^7/_2]_3$
2. $3d\,^7(^4P)4s4p(^3P^\circ)$ D° $^3[^5/_2]^\circ_{7/2}$

The orbital angular momentum of the core is coupled with the orbital angular momentum of the external electron(s) to give the total orbital angular momentum L. The letter symbol for the final L value is listed with the configuration because this angular momentum is then coupled with the spin of the core (S_1) to obtain the resultant K angular momentum of the final term (in brackets). The multiplicity of the [K] term arises from the spin of the external electron(s).

Coupling Schemes and Term Symbols

The coupling schemes outlined above include those now most frequently used in calculations of atomic structure. Any term symbol gives the values of two angular momenta that may be coupled to give the total electronic angular momentum of a level (indicated by the J value). For configurations of more than one unfilled subshell, the angular momenta involved in the final coupling derive from two groups of electrons (either group may consist of only one electron). These are often an inner group of coupled electrons and an outer group of coupled electrons, respectively. In any case the quantum numbers for the two groups can be distinguished by subscripts 1 and 2, so that quantum numbers represented by capital letters without subscripts are total quantum numbers for both groups. Thus, the quantum numbers for the two vectors that couple to give the final J are related to the term symbol as follows:

Coupling Scheme	Quantum numbers for vectors that couple to give J	Term Symbol
LS	L,S	^{2S+1}L
$J_1 J_2$	J_1, J_2	(J_1, J_2)
$J_1 L_2(\rightarrow K)$	K, S_2	$2S_2$+1[K]
$LS_1(\rightarrow K)$	K, S_2	$2S_2$+1[K]

The parity is indicated by appended degree symbols on odd parity terms.

EIGENVECTOR COMPOSITION OF LEVELS

The wave functions of levels are often expressed as eigenvectors that are linear combinations of basis states in one of the standard coupling schemes. Thus, the wave function $\Psi(\alpha J)$ for a level labeled αJ might be expressed in terms of normalized *LS* coupling basis states $\Phi(\gamma SLJ)$:

$$\psi(\alpha J) = \sum_{\gamma SL} c(\gamma SLJ)\Phi(\gamma SLJ)$$

The $c(\gamma SLJ)$ are expansion coefficients, and

$$\sum_{\gamma SL} \left|c(\gamma SLJ)\right|^2 = 1$$

The squared expansion coefficients for the various γSL terms in the composition of the αJ level are conveniently expressed as percentages, whose sum is 100 %. Thus the percentage contributed by the pure Russell-Saunders state γSLJ is equal to $100 \times |c(\gamma SLJ)|^2$. The notation for Russell-Saunders basis states has been used only for concreteness; the eigenvectors may be expressed in any coupling scheme, and the coupling schemes may be different for different configurations included in a single calculation (with configuration interaction). "Intermediate coupling" conditions for a configuration are such that calculations in both *LS* and *jj* coupling yield some eigenvectors representing significant mixtures of basis states.

The largest percentage in the composition of a level is called the *purity* of the level in that coupling scheme. The coupling scheme (or combination of coupling schemes if more than one configuration is involved) that results in the largest average purity for all the levels in a calculation is usually best for naming the levels. With regard to any particular calculation, one does well to remember that, as with other calculated quantities, the resulting eigenvectors depend on a specific theoretical model and are subject to the inaccuracies of whatever approximations the model involves.

Theoretical calculations of experimental energy level structures have yielded many eigenvectors having significantly less than 50 % purity in any coupling scheme. Since many of the corresponding levels have nevertheless been assigned names by spectroscopists, some caution is advisable in the acceptance of level designations.

GROUND LEVELS AND IONIZATION ENERGIES FOR THE NEUTRAL ATOMS

The ground-state electron configurations of elements heavier than neon are shortened in the table by using rare-gas element symbols in brackets to represent the corresponding electrons. The ground levels of all neutral atoms have reasonably meaningful *LS*-coupling names, the corresponding eigenvector percentages lying in the range from ~55 % to 100 %. These names are listed in the table, except for Pa, U, and Np; the lowest few ground-configuration levels of these atoms comprise better $5f^N(L_1S_1J_1)$, $6d_j7s^2$ (J_1j) terms than *LS*-coupling terms. The relatively large spin-orbit interaction of the $6p$ electrons produces *jj*-coupling structures for the $(6p^2{}_{1/2})_0$, $(6p^2{}_{1/2}6p_{3/2})^o{}_{3/2}$, and $(6p^2{}_{1/2}6p^2{}_{3/2})_2$ ground levels of the $6p^2$, $6p^3$, and $6p^4$ configurations of neutral Pb, Bi, and Po, respectively. As noted in the section *jj* Coupling of Equivalent Electrons, the *jj*-coupling names are more appropriate for these atoms than the alternative *LS*-coupling designations in the table.

The ionization energies in the table are from recent compilations. The uncertainties are mainly in the range from less than one to several units in the last decimal place, but a few of the values may be in error by 20 or more units in the final place; i.e., the error of some of the two place values could be greater than 0.2 eV. Although no more than four decimal places are given here, values for both the neutral and singly-ionized atoms are given to their full accuracies in.

ZEEMAN EFFECT

The Zeeman effect for "weak" magnetic fields (the anomalous Zeeman effect) is of special interest because of the importance of Zeeman data in the analysis and theoretical interpretation of complex spectra. In a weak field, the J value remains a good quantum number although in general a level is split into magnetic sublevels. The g value of such a level may be defined by the expression for the energy shift of its magnetic sublevel having magnetic quantum number M, which has one of the $2J + 1$ values, $-J, -J + 1, ..., J$:

$$\Delta E = gM\, \mu_B B$$

The magnetic flux density is B, and μ_B is the Bohr magneton ($\mu_B = e\hbar/2m_e$).

The wavenumber shift $\Delta\sigma$ corresponding to this energy shift is

$$\Delta\sigma = gM\, (0.466\,86\; B\ \text{cm}^{-1})\ ,$$

with B representing the numerical value of the magnetic flux density in teslas. The quantity in parentheses, the Lorentz unit, is of the order of 1 or 2 cm^{-1} for typical flux densities used to obtain Zeeman-effect data with classical spectroscopic methods. Accurate data can be obtained with much smaller fields, of course, by using higher-resolution techniques such as laser spectroscopy. Most of the g values now available for atomic energy levels were derived by application of the above formula (for each of the two combining levels) to measurements of optical Zeeman patterns. A single transverse-Zeeman-effect pattern (two polarizations, resolved components, and sufficiently complete) can yield the J value and the g value for each of the two levels involved.

Neglecting a number of higher-order effects, we can evaluate the g value of a level βJ belonging to a pure LS-coupling term using the formula

$$g_{\beta SLJ} = 1+\left(g_{e-1}\right)\frac{J(J+1)-L(L+1)+S(S+1)}{2J(J+1)}$$

The independence of this expression from any other quantum numbers (represented by β) such as the configuration, etc., is important. The expression is derived from vector coupling formulas by assuming a g value of unity for a pure orbital angular momentum and writing the g value for a pure electron spin as g_e. A value of 2 for g_e yields the Landé formula. If the anomalous magnetic moment of the electron is taken into account, the value of g_e is 2.002 319 3. "Schwinger" g values obtained with this more accurate value for g_e are given for levels of *SL* terms in Ref..

The usefulness of g_{SLJ} values is enhanced by their relation to the g values in intermediate coupling. In the notation used in Equation for the wave function of a level βJ in intermediate coupling, the corresponding g value is given by

$$g_{\beta J} = \sum_{\gamma SL} g_{SLJ} \left|c(\gamma SLJ)\right|^2$$

where the summation is over the same set of quantum numbers as for the wave function. The $g_{\beta J}$ value is thus a weighted average of the Landé g_{SLJ} values, the weighting factors being just the corresponding component percentages.

Formulas for magnetic splitting factors in the $J_1 J_2$ and $J_1 L_2$ coupling schemes are given in Refs. and. Some higher-order effects that must be included in more accurate Zeeman-effect calculations are treated by Bethe and Salpeter and by Wybourne, for example. High precision calculations for helium are given in Ref..

TERM SERIES, QUANTUM DEFECTS, AND SPECTRAL-LINE SERIES

The Bohr energy levels for hydrogen or for a hydrogenic (one-electron) ion are given by

$$E_n = -\frac{Z^2}{n^2},$$

in units of R_M, the Rydberg for the appropriate nuclear mass. For a multielectron atom, the deviations of a series of (core) *nl* levels from hydrogenic E_n values may be due mainly to core penetration by the *nl* electron (low *l*-value series), or core polarization by the *nl* electron (high *l*-value series), or a combination of the two effects. In either case it can be shown that these deviations can be approximately represented by a constant quantum defect δ_l in the Rydberg formula,

$$E_{nl} = \frac{Z_c^2}{(n-\delta_l)^2} = -\frac{Z_c^2}{(n*)^2}$$

where Z_c is the charge of the core and $n^* = n - \delta$ is the effective principal quantum number. If the core includes only closed subshells, the E_{nl} values are with respect to a value of zero for the (core)$^1S_0$ level, i.e., the $^1S_0$ level is the limit of the (core)*nl* series. If the quantities in equation are taken as positive, they represent term values or ionization energies; the term value of the ground level of an atom or ion with respect to the ground level of the next higher ion is thus the principal ionization energy.

If the core has one or more open subshells, the series limit may be the baricenter of the entire core configuration, or any appropriate sub-structure of the core, down to and including a single level. The E_{nl} values refer to the series of corresponding (core)*nl* structures built on the particular limit structure. The value of the quantum defect depends to some extent on which (core)*nl* structures are represented by the series formula.

The quantum defect in general also has an energy dependence that must be taken into account if lower members of a series are to be accurately represented by Equation. For an unperturbed series, this dependence can be expressed by the extended Ritz formula

$$\delta = n - n^*$$

$$= \delta_0 + \frac{a}{(n-\delta_0)^2} + \frac{b}{(n-\delta_0)^4} + \ldots,$$

with $\delta_0, a, b\ldots$ constants for the series (δ_0 being the limit value of the quantum defect for high series members). The value of *a* is usually positive for core-penetration series and negative for core-polarization series. A discussion of the foundations of the Ritz expansion and application to high precision calculations in helium is given in Ref..

A spectral-line series results from either emission or absorption transitions involving a common lower level and a series of successive (core)*nl* upper levels differing only in their *n* values. The principal series of Na I, $3s\ ^2S_{1/2} - np\ ^2P^\circ_{1/2,3/}$

$_2$ (n ≥ 3) is an example. The regularity of successive upper term values with increasing n is of course observed in line series; the intervals between successive lines decrease in a regular manner towards higher wavenumbers, and the series of increasing wavenumbers converges towards the term value of the lower level as a limit.

SEQUENCES

Several types of sequences of elements and/or ionization stages are useful because of regularities in the progressive values of parameters relating to structure and other properties along the sequences. All sequence names may refer either to the atoms and/or ions of the sequence or to their spectra.

Isoelectronic Sequence

A neutral atom and those ions of other elements having the same number of electrons as the atom comprise an isoelectronic sequence. An isoelectronic sequence is named according to its neutral member; for example, the Na I isolectronic sequence.

Isoionic, Isonuclear, and Homologous Sequences

An isoionic sequence comprises atoms or ions of different elements having the same charge. Such sequences have probably been most useful along the *d*- and *f*-shell rows of the periodic table. Isoionic analyses have also been carried out along *p*-shell rows, however, and a fine-structure regularity covering spectra of the *p*-shell atoms throughout the periodic table is known.

The atom and successive ions of a particular element comprise the isonuclear sequence for that element.

The elements of a particular column and subgroup in the periodic table are homologous. Thus the C, Si, Ge, Sn, and Pb atoms belong to a homologous sequence having np^2 ground configurations. The singly ionized atoms of these elements comprise another example of a homologous sequence.

SPECTRAL WAVELENGTH RANGES, DISPERSION OF AIR

The ranges of most interest for optical atomic spectroscopy are:

~ 2 - 20 μm	mid-infrared (ir)
700 - 2000 nm	near ir
400 - 700 nm	visible
200 - 400 nm	near ultraviolet (uv)
100 - 200	vacuum uv or far uv
10 - 100 nm	extreme uv (euv or xuv)
< 10 nm	soft x-ray, x-ray

The above correspondence of names to ranges should not be taken as exact; the variation as to the extent of some of the named ranges found in the literature is considerable.

Wavelengths in standard air are often tabulated for the region longer than 200 nm. These wavelengths can be related to energy-level differences by conversion to the corresponding (vacuum) wavenumbers or frequencies.

WAVELENGTH (FREQUENCY) STANDARDS

In 2001 the Comité International des Poids et Mesures recommended values for optical frequency standards from stabilized lasers using various absorbing atoms, atomic ions, and molecules. These frequencies range from 29 054 057 446 579 Hz (10.318 436 884 460 μm, relative standard uncertainty 1.4×10^{-13} for a transition in OsO_4 to 1 267 402 452 889.92 kHz (236.540 853 549 75 nm; relative standard uncertainty 3.6×10^{-13}) for a transition in the $^{115}In^+$ ion.

Extensive tables of wavenumbers for molecular transitions in the mid-ir range 2.3 to 20.5 μm are included in a calibration atlas published in 1991. Wavenumbers of Ar I and Ar II emission lines having uncertainties as small as 0.0003 cm^{-1} are included in tables for these spectra covering a broad range from 222 nm to 5.865 μm. Measurements of U and Th lines (575 nm to 692 nm) suitable for wavenumber calibration at uncertainty levels of 0.0003 cm^{-1} or 0.0004$ cm^{-1} were reported in Ref.. Comprehensive tables of lines for U, Th, and I_2 are useful for calibration at uncertainty levels of 0.002 cm^{-1} to 0.003 cm^{-1}, the atlas of the Th spectrum extending down to 278 nm.

A 1974 compilation gives reference wavelengths for some 5400 lines of 38 elements covering the range 1.5 nm to 2.5 μm, with most uncertainties between 10^{-5} and 2×10^{-4} nm. The wavelengths for some 1100 Fe lines selected from the Fe/Ne hollow-cathode spectrum have been recommended for reference standards over the range 183 nm to 4.2 μm, with wavenumber uncertainties 0.001 cm^{-1} to 0.002 cm^{-1}. Wavelengths for about 3000 vuv and uv lines (110 nm to 400 nm) from a Pt/Ne hollow-cathode lamp have been determined with uncertainties of 0.0002 nm or less. More recent high-accuracy measurements of ultraviolet lines of Fe I, Ge I, Kr II, and Pt I, II, include some wavelengths with uncertainties smaller than 10^{-5} nm. The wavelengths tabulated for the Kr and Pt lines in extend from 171 nm to 315 nm, and the accuracies of earlier measurements of a number of spectra useful for wavelength calibration are discussed.

SPECTRAL LINES: SELECTION RULES, INTENSITIES, TRANSITION PROBABILITIES, VALUES, AND LINE STRENGTHS

Emission Intensities (Transition Probabilities)

The total power ε radiated in a spectral line of frequency ν per unit source volume and per unit solid angle is

$$\epsilon_{line} = (4\pi)^{-1} h^{\nu} A_{ki} N_k$$

where A_{ki} is the atomic transition probability and N_k the number per unit volume (number density) of excited atoms in the upper (initial) level *k*. For a homogeneous light source of length *l* and for the optically thin case, where all radiation escapes, the total emitted line intensity (SI quantity: radiance) is

$$I_{line} = \epsilon_{line}\, l = \mathrm{int}_0^{+\infty}\, I(\lambda)\, d\lambda \;\; (4\pi)^{-1} (hc / \lambda_0)\, A_{ki} N_k l$$

where $I(\lambda)$ is the specific intensity at wavelength λ, and λ_0 the wavelength at line center.

Absorption *f* Values

In absorption, the reduced absorption

$$W(\lambda)=[I(\lambda)-I'(\lambda)]/I(\lambda)$$

is used, where $I(\lambda)$ is the incident intensity at wavelength &ambda;, e.g., from a source providing a continuous background, and $I'(\lambda)$ the intensity after passage through the absorbing medium. The reduced line intensity from a homogeneous and optically thin absorbing medium of length l follows as

$$W_{ik} = \int_0^{+\infty} W(\lambda)\,d\lambda = \frac{e^2}{4\in_0 m_e c^2}\lambda_0^2 N_i f_{ik} l$$

where f_{ik} is the atomic (absorption) oscillator strength (dimensionless).

Line Strengths

A_{ki} and f_{ik} are the principal atomic quantities related to line intensities. In theoretical work, the *line strength S* is also widely used:2

$$S = S(i,k) = S(k,i) = |R_{ik}|^2$$

$$R_{ik} = \langle \psi_k | P | \psi_i \rangle,$$

where ψ_i and ψ_k are the initial- and final-state wave functions and R_{ik} is the *transition matrix element* of the appropriate multipole operator P (R_{ik} involves an integration over spatial and spin coordinates of all N electrons of the atom or ion).

***Relationships between* A, f, *and* S**

The relationships between *A*, *f*, and *S* for electric dipole (E1, or allowed) transitions in SI units (*A* in s^{-1}, λ in m, *S* in $m^2 C^2$) are

$$A_{ki} = \frac{2\pi e^2}{m_e c \in_0 \lambda^2}\frac{gi}{gk} f_{ik} = \frac{16\pi^3}{3h \in_0 \lambda^3 gk} S$$

Numerically, in customary units (A in s^{-1}, λ in Å, S in atomic units),

$$A_{ki} = \frac{6.6702\times10^{15}}{\lambda^2}\frac{gi}{gk}f_{ik} = \frac{2.0261\times10^{18}}{\lambda^3 gk}S$$

and for S and ΔE in atomic units,

$$f_{ik} = \frac{2}{3}(\Delta E / g_i)S$$

g_i and g_k are the statistical weights, which are obtained from the appropriate angular momentum quantum numbers. Thus for the lower (upper) level of a spectral line, $g_{i(k)} = 2J_{i(k)} + 1$, and for the lower (upper) term of a multiplet,

$$\bar{g}_i(k) = \sum_{i(k)}\left(2J_{i(k)}+1\right)$$

$$= \left(2L_{i(k)}+1\right)\left(2S_{i(k)}+1\right).$$

The A_{ki} values for strong lines of selected elements are given. For comprehensive numerical tables of A, f, and S, including forbidden lines.

Conversion relations between S and A_{ki} for the most common forbidden transitions

	SI units	Numerically, in customary units
Electric quadrupole	$A_{ki} = \frac{16\pi^5}{15h\in 0\lambda^5 gk}S$	$A_{ki} = \frac{1.1199\times10^{18}}{gk\lambda^5}S$
Magnetic dipole	$A_{ki} = \frac{16\pi^3\mu_0}{3h\lambda^3 gk}S$	$A_{ki} = \frac{2.697\times10^{13}}{gk\lambda^3}S$

A in s^{-1}, λ in m. Electric quadrupole: S in $m^4\ C^2$. Magnetic dipole: S in $J^2\ T^{-2}$.

A in s^{-1}, λ in Å. S in atomic units: $= 2.013 \times 10^{-79}$ m^4 C^2 (electric quadrupole), $e^2h^2/16\pi^2m_e^2 = \mu_B^2 = 8.601 \times 10^{-47}$ J^2 T^{-2} (magnetic dipole). μ_B is the Bohr magneton.

Oscillator strengths *f* are not used for forbidden transitions, i.e., magnetic dipole (M_1), electric quadrupole (E_2), etc.

Relationships between Line and Multiplet Values

The relations between the total strength and *f* value of a multiplet (M) and the corresponding quantities for the lines of the multiplet (allowed transitions) are

$$S_M = \sum S_{line}$$

$$f_M = \left(\bar{\lambda}\bar{g}_i\right)^{-1} \sum_{j_k, J_i} g_i \lambda(J_i, J_k) f(J_i, J_k)$$

$\bar{\lambda}$ is the weighted ("multiplet") wavelength in vacuum:

$$\bar{\lambda} = n\bar{\lambda}_{air} = hc/\overline{\Delta E},$$

where

$$\overline{\Delta E} = \overline{E_k} - \overline{E_i} = \left(\bar{g}_k\right)^{-1} \sum_{J_k} g_k E_k - \left(\bar{g}_i\right)^{-1} \sum_{J_i} g_i E_i$$

and *n* is the refractive index of standard air.

SPECTRAL LINES: SELECTION RULES, INTENSITIES, TRANSITION PROBABILITIES, VALUES, AND LINE STRENGTHS

Relative Strengths for Lines of Multiplets in *LS* Coupling

This table lists relative line strengths for frequently encountered symmetrical (P → P, D → D) and normal (S → P, P → D) multiplets in *LS* coupling. The strongest, or principal, lines are situated along the main diagonal of the table and are called x_1, x_2, etc. Their strengths normally diminish along the diagonal. The satellite lines y_n and z_n are usually weaker and deviate more from the *LS* values than the stronger diagonal lines when departures from *LS* coupling are encountered.

ATOMIC LIFETIMES

The radiative lifetime τ_k of an atomic level k is related to the sum of transition probabilities to all levels i lower in energy than k:

$$\tau_k = \left(\sum_i A_{ki}\right)^{-1}$$

The *branching* ratio of a particular transition, say to state i', is defined as

$$A_{ki'} / \sum_i A_{ki} = A_{ki} / \tau_k$$

If only one branch (i') exists (or if all other branches may be neglected), one obtains $A_{ki'}\ \tau_k = 1$, and

$$\tau_k = 1 / A_{ki'} \tau_k$$

REGULARITIES AND SCALING

Transitions in Hydrogenic (One-Electron) Species

The nonrelativistic *energy* of a hydrogenic transition is

$$(\Delta E)_z = (E_k - E_i)_Z = R_M hcZ^2 \left(1/n_i^2 - 1/n_k^2\right)$$

Hydrogenic Z scaling. The spectroscopic quantities for a hydrogenic ion of nuclear charge Z are related to the equivalent quantities in hydrogen (Z = 1) as follows (neglecting small differences in the values of R_M):

$$(\Delta E)_z = Z^2 (\Delta E)_H$$

$$(\lambda_{vac})_z = Z^2 (\lambda_{vac})_H$$

$$S_Z = Z^{-2} S_H$$

$$f_Z = f_H$$

$$A_Z = Z^4 A_H$$

For large values of Z, roughly $Z > 20$, relativistic corrections become noticeable and must be taken into account.

f-value trends. f values for high series members (large *n*2 values) of hydrogenic ions decrease according to

$$f(n,l \rightarrow n',l \pm 1) \alpha (n')^{-3}$$

Data for some lines of the main spectral series of hydrogen are given in the table below.

Some transitions of the main spectral series of hydrogen

Transition	Customary name	λ (Å)	g_i	g_k	A_{ki} (10^8 s^{-1})
1-2	(L_α)	1 215. 67	2	8	4.699
1-3	(L_β)	1 025. 73	2	18	5.575(–1)
1-4	(L_γ)	972. 537	2	32	1.278(–1)
1-5	(L_δ)	949. 743	2	50	4.125(–2)
1-6	(L_ε)	937. 80	2	72	1.644(–2)
2-3	(H_α)	6 562. 80	8	18	4.410(–1)
2-4	(H_β)	4 861. 32	8	32	8.419(–2)
2-5	(H_γ)	4 340. 46	8	50	2.530(–2)
2-6	(H_δ)	4 101. 73	8	72	9.732(–3)
2-7	(H_ε)	3 970. 07	8	98	4.389(–3)
3-4	(P_α)	18 751. 0	18	32	8.986(–2)
3-5	(P_β)	12 818. 1	18	50	2.201(–2)
3-6	(P_γ)	10 938. 1	18	72	7.783(–3)
3-7	(P_δ)	10 049. 4	18	98	3.358(–3)
3-8	(P_ε)	9 545. 97	18	128	1.651(–3)

L_α is often called Lyman α, H_α = Balmer α, P_α = Paschen α, etc.

Wavelengths below 2000 Å are in vacuum; values above 2000 Å are in air. For transitions in hydrogen, $g_{i(k)} = 2(n_{i(k)})^2$, where $n_{i(k)}$, is the principal quantum number of the lower (upper) electron shell.

The number in parentheses indicates the power of 10 by which the value has to be multiplied.

Systematic Trends and Regularities in Atoms and Ions with Two or more Electrons

Nonrelativistic atomic quantities for a given state or transition in an *isoelectronic sequence* may be expressed as power series expansions in Z^{-1}:

$$Z^{-2} E = E_0 + E_1 Z^{-1} + E_2 Z^{-2} + \ldots ,$$

$$Z^2 S = S_0 + S_1 Z^{-1} + S_2 Z^{-2} + \ldots ,$$

$$f = f_0 + f_1 Z^{-1} + f_2 Z^{-2} + \ldots ,$$

where E_0, f_0, and S_0 are hydrogenic quantities. For transitions in which n does not change ($n_i = n_k$), $f_0 = 0$, since states i and k are degenerate. For equivalent transitions of *homologous atoms*, f values vary gradually.

$$(nl - n'l')_{Li} \rightarrow [(n+1)l - (n'+1)l']_{Na}$$

$$\rightarrow [(n+2)l - (n'+2)l']_{Cu} \rightarrow \ldots.$$

Complex atomic structures, as well as cases involving strong cancellation in the integrand of the transition integral, generally do not adhere to this regular behaviour.

SPECTRAL LINE SHAPES, WIDTHS, AND SHIFTS

Observed spectral lines are always broadened, partly due to the finite resolution of the spectrometer and partly due to intrinsic physical causes. The principal physical causes of spectral line broadening are Doppler and pressure broadening.

Doppler Broadening

Doppler broadening is due to the thermal motion of the emitting atoms or ions. For a Maxwellian velocity distribution, the line shape is *Gaussian*; the full width at half maximum intensity (FWHM) is, in Å,

$$\Delta\lambda_{1/2}^{D} = \left(7.16\times10^{-7}\right)\lambda\left(T/M\right)^{1/2}$$

T is the temperature of the emitters in K, and M the atomic weight in atomic mass units (amu).

Pressure Broadening

Pressure broadening is due to collisions of the emitters with neighbouring particles. Shapes are often approximately Lorentzian, i.e., $I(\lambda)$ $\{1 + [(\lambda - \lambda_0)/\Delta\lambda_{1/2}]^2\}^{-1}$. In the following formulas, all FWHM's and wavelengths are expressed in Å, particle densities N in cm^{-3}, temperatures T in K, and energies E or I in cm^{-1}.

Resonance broadening (self-broadening) occurs only between identical species and is confined to lines with the upper or lower level having an electric dipole transition (resonance line) to the ground state. The FWHM may be estimated as

$$\Delta\lambda_{1/2}^{R} \simeq 8.6\times10^{-30}\,(gi/gk)^{1/2}\,\lambda^{2}\lambda_{r}f_{r}N_{i}$$

where λ is the wavelength of the observed line. f_r and λ_r are the oscillator strength and wavelength of the resonance line; g_k and g_i are the statistical weights of its upper and lower levels. N_i is the ground state number density.

For the $1s2p\ {}^1\mathrm{P} - 1s3d\ {}^1\mathrm{D}_2$ transition in He I [$\lambda = 6678.15$ Å; $\lambda_r\,(1s^2\ {}^1\mathrm{S}_0 - 1s2p\ {}^1\mathrm{P}) = 584.334$ Å; $g_i = 1$; $g_k = 3$; $f_r = 0.2762$] at $N_i = 1 \times 10^{18}$ cm^{-3}: $\Delta\lambda^{R}_{1/2} = 0.036$ Å.

Van der Waals broadening arises from the dipole interaction of an excited atom with the induced dipole of a ground state atom. (In the case of foreign gas broadening, both the perturber and the radiator may be in their respective ground states.) An approximate formula for the FWHM, strictly applicable to hydrogen and similar atomic structures only, is

$$\Delta\lambda_{1/2}^{W} \simeq 30\times10^{16}\,\lambda^{2}C_{6}^{2/5}\,(T/\mu)^{3/10}\,N$$

where μ is the atom-perturber reduced mass in units of u, N the perturber density, and C_6 the interaction constant. C_6 may be roughly estimated as follows: $C_6 = C_k - C_i$, with $C_{i\,(k)} = (9.8 \times 10^{10})$ (α_d $R^2_i\,(k)$ α_d in cm^3, R^2 in a_0^2). Mean atomic polarizability $\alpha_d \approx (6.7 \times 10^{-25})$ $(3I_H/4E^*;)^2$ cm^3, where I_H is the ionization energy of hydrogen and E^*; the energy of the first excited level of the perturber atom. $R^2_{i\,(k)} \approx 2.5\,[I_H/(I-E_{i\,(k)})]^2$, where I is the ionization energy of the radiator. Van der Waals broadened lines are red shifted by about one-third the size of the FWHM.

For the $1s2p\ {}^1\mathrm{P}^{\circ}_1 - 1s3d\ {}^1\mathrm{D}_2$ transition in He I, and with He as perturber: $\lambda = 6678.15$ Å; $I = 198\,311$ cm^{-1}; $E^* = E_i = 171\,135$ cm^{-1}; $E_k = 186\,105$ cm^{-1}; $\mu = 2$. At $T = 15\,000$ K and $N = 1 \times 10^{18}$ cm^{-3}: $\Delta\lambda^{W}_{1/2} = 0.044$ Å.

Stark broadening due to charged perturbers, i.e., ions and electrons, usually dominates resonance and van der Waals broadening in discharges and plasmas. The FWHM for hydrogen lines is

$$\Delta\lambda_{1/2}^{S,H} = \left(2.50\times10^{-9}\right)\alpha_{1/2}N_e^{2/3},$$

where N_e is the electron density. The half-width parameter $\alpha_{1/2}$ for the H_β line at 4861 Å, widely used for plasma diagnostics, is tabulated in the table below for some typical temperatures and electron densities. This reference also contains $\alpha_{1/2}$ parameters for other hydrogen lines, as well as Stark width and shift data for numerous lines of other elements, i.e., neutral atoms and singly charged ions (in the latter, Stark widths and shifts depend linearly on N_e). Other tabulations of complete hydrogen Stark profiles exist.

Values of Stark-broadening parameter $\alpha_{1/2}$ for the H_β line of hydrogen (4861 Å) for various temperatures and electron densities.

	N_e (cm^{-3})			
T(K)	**10^{15}**	**10^{16}**	**10^{17}**	**10^{18}**
5 000	0.0787	0.0808	0.0765	...
10 000	0.0803	0.0840	0.0851	0.0781
20 000	0.0815	0.0860	0.0902	0.0896
30 000	0.0814	0.0860	0.0919	0.0946

SPECTRAL CONTINUUM RADIATION

Hydrogenic Species

Precise quantum-mechanical calculations exist only for hydrogenic species. The total power ε_{cont} radiated (per unit source volume and per unit solid angle, and expressed in SI units) in the wavelength interval $\Delta\lambda$ is the sum of radiation due to the recombination of a free electron with a bare ion (free-bound transitions) and bremsstrahlung (free-free transitions):

$$\epsilon_{\text{cont}} = \frac{\epsilon^6}{2\pi\,\epsilon_0^3\left(6\pi m_e\right)^{3/2}} N_e N_Z Z^2 \times$$

$$\frac{1}{(kT)^{1/2}} \exp\left(-\frac{hc}{\lambda kT}\right)\frac{\Delta\lambda}{\lambda^2}$$

$$\times\left\{\frac{2Z^2 I_H}{kT} \sum_{n \geq (z^2 I_H \lambda / hc)^{1/2}}^{n'} \frac{\gamma fb}{n^3} \exp\left(\frac{z^2 I_H}{n^2 kT}\right)\right.$$

$$+\bar{\gamma} fb \left[\exp\left(\frac{Z^2 I_H}{(n'+1)^2 kt}\right) - 1\right]$$

where N_e is the electron density, N_Z the number density of hydrogenic (bare) ions of nuclear charge Z, I_H the ionization energy of hydrogen, $n2$ the principal quantum number of the lowest level for which adjacent levels are so close that they approach a continuum and summation over n may be replaced by an integral. (The choice of n' is rather arbitrary; n' as low as 6 is found in the literature.) γ_{fb} and γ_{ff} are the Gaunt factors, which are generally close to unity. (For the higher free-bound continua, starting with $n' + 1$, an average Gaunt factor $\bar{\gamma}_{fb}$ is used.) For neutral hydrogen, the recombination continuum forming H^- becomes important.

In the equation above, the value of the constant factor is 6.065×10^{-55} W m^4 J$^{1/2}$ sr^{-1}. [Numerical example: For atomic hydrogen ($Z = 1$), the quantity ε_{cont} has the value 2.9 W m^{-3} sr^{-1} under the following conditions: $\lambda = 3 \times 10^{-7}$ m; $\Delta\lambda = 1 \times 10^{-10}$ m; N_e ($= N_{Z=1}$) $= 1 \times 10^{21}$ m^{-3}; $T = 12\,000$ K. The lower limit of the summation index n is 2; the upper limit $n2$ has been taken to be 10. All Gaunt factors γ_{fb}, $\bar{\gamma}_{f\ b}$, γ_{ff} have been assumed to be unity.]

Many-Electron Systems

For many-electron systems, only approximate theoretical treatments exist, based on the quantum-defect method. Experimental work is centered on the noble gases. Modifications of the continuum by autoionization processes must also be considered.

Near the ionization limit, the f values for bound-bound transitions of a spectral series ($n' \to \infty$) make a smooth connection to the differential oscillator strength distribution $df/d\varepsilon$ in the continuum.

Chapter 3

Molecular Modeling

QUANTUM CHEMICAL METHODS

Aspects of molecular structure and dynamics can be modeled using classical methods in the form of molecular mechanics and dynamics. The classical force field is based on empirical results, averaged over a large number of molecules. Because of this extensive averaging, the results can be good for standard systems, but there are many important questions in chemistry that can not at all be addressed by means of this empirical approach. If one wants to know more than just structure or other properties that are derived only from the potential energy surface, in particular properties that depend directly on the electron density distribution, one has to resort to a more fundamental and general approach: quantum chemistry. The same holds for all non-standard cases for which molecular mechanics is simply not applicable.

Quantum chemistry is based on the postulates of Quantum Mechanics. The theory of quantum chemistry with an emphasis on their practical implications for the molecular modeler, and we will try to answer the questions of when to use a quantum chemical method instead of molecular mechanics, which quantum chemical method to choose, and what to expect from the quality of the results.

In quantum chemistry, the system is described by a wavefunction which can be found by solving the Schrödinger equation. This equation relates the stationary states of the system and their energies to the Hamiltonian operator, which can be viewed as the recipe for obtaining the energy associated with a wavefunction describing the positions of the nuclei and electrons in the system. In practice the Schršdinger equation cannot be solved exactly and approximations have to be made, as we shall see below.

The approach is called "ab initio" when it makes no use of empirical information, except for the fundamental constants of nature such as the mass of the electron, Planck's constant etc., that are required to arrive at numerical predictions. Do not confuse "ab initio" with "exact" ! In spite of the necessary approximations, ab initio theory has the conceptual advantage of generality, and the practical advantage that (with experience) its successes and failures are more or less predictable.

The major disadvantage of ab initio quantum chemistry are the heavy demands on computer power. Therefore, further approximations have been applied for a long time which go together with the introduction of empirical parameters into the theoretical model. This has led to a number of semi-empirical quantum chemical methods, which can be applied to larger systems, and give reasonable electronic wavefunctions so that electronic properties can be predicted. Compared with ab initio calculations their reliability is less and their applicability is limited by the requirement for parameters, just like in molecular mechanics.

In general, one should apply quantum chemistry for "small" systems, which can be treated at a very high level, when electronic properties are sought (electric moments, polarizabilities, shielding constants in NMR and ESR, etc.) and for "non-standard" structures, for which no valid molecular mechanics parameters are available. Examples are conjugated pi systems, organometallic compounds and other systems with unusual bond or atom types, excited states, reactive intermediates, and generally structures with unusual electronic effects.

THE SCHRODINGER EQUATION

The energies and wavefunctions of stationary states of a system are given by the solutions of the Schrödinger Equation: $\hat{H}\psi = E\psi$

In this equation $\hat{H}$ is the Hamiltonian operator which in this case gives the kinetic and potential energies of a system of atomic nuclei and electrons. As we shall see below it is analogous to the classical kinetic energy of the particles and the Coulomb electrostatic interactions between the nuclei and electrons. ψ is a wavefunction, one of the solutions of the eigenvalue equation. This wavefunction depends on the coordinates of the electrons and the nuclei. The Hamiltonian is composed of three parts: the kinetic energy of the nuclei, the kinetic energy of the electrons, and the potential energy of nuclei and electrons.

Schrödinger equation: $\hat{H}\psi_{e,n} = E\psi_{e,n}$

Hamiltonian: $\hat{H} = \hat{T}_n + \hat{T}_e + \hat{V}_{e,n}$

Four approximations are commonly (but not necessarily) made:

- Time independence; we are looking at states that are stationary in time.
- Neglect of relativistic effects; this is warranted unless the velocity of the electrons approaches the speed of light, which is the case only in heavy atoms with very high nuclear charge.
- Born-Oppenheimer approximation; separation of the motion of nuclei and electrons.
- Orbital approximation; the electrons are confined to certain regions of space.

The Born-Oppenheimer approximation implies the separation of nuclear and electronic wavefunctions, the total wavefunction being a product of the two:

Born-Oppenheimer: $\psi_{e,n} = \chi_n \psi_e$

The motivation behind this is that the electrons are so much lighter than the nuclei that their motion can easily follow the nuclear motion. In practice, this approximation is usually valid. From this point we will look at the electronic wavefunction ψ_e which is obtained by solving the *electronic Schrödinger equation*:

$$\hat{H}_e(R_n)\psi(r_e) = E_e(R_n)\psi(r_e)$$

This equation still contains the positions of the nuclei, however not as variables but as parameters.

The electronic Hamiltonian contains three terms: kinetic energy, electrostatic interaction between electrons and nuclei, and electrostatic repulsion between electrons. In order to simplify expressions and to make the theory independent of the experimental values of physical constants, atomic units are introduced:

$e = 1$ charge of electron

$m = 1$ mass of the electron

$\hbar = 1$ Planck's constant divided by 2 pi

Derived atomic units of length and energy are:

$$1 \text{ bohr} = a_0 = \frac{\hbar^2}{me^2} = 0.529\text{A}$$

$$1 \text{ hartree} = \frac{e^2}{a_0} = 4.3598 \times 10^{-18} \text{ J} = 627.51 \text{ kcal/mol}$$

With these units the electronic Hamiltonian is:

$$\hat{H}_e = -\frac{1}{2}\sum_{i=1}^{n}\Delta_i - \sum_{i=1}^{n}\sum_{A=1}^{N}\frac{Z_A}{|R_A - r_i|} + \sum_{i<j}^{n}\frac{1}{r_{ij}}$$

The Δ symbol is the Laplace operator $\Delta_i = \frac{\partial^2}{\partial x_i^2} + \frac{\partial^2}{\partial y_i^2} + \frac{\partial^2}{\partial z_i^2}$ (also called "del-squared"). The total energy in the Born-Oppenheimer model is obtained by adding the nuclear repulsion energy to the electronic energy:

$$E_{tot} = E_e + E_n$$

$$E_n = \sum_{A<B}^{N} \frac{Z_A Z_B}{|R_A - R_B|}$$

The total energy defines a potential energy hypersurface $E = f(Q)$ which can be used to subsequently solve a Schrödinger equation for the nuclear motion:

$$\left[\hat{T}_n + E(R_n)\right]\Phi(R_n) = \varepsilon_i \Phi(R_n)$$

In the following section we shall deal with the important problem of solving the electronic Schrödinger equation.

SOLVING THE ELECTRONIC SCHRÖDINGER EQUATION: HARTREE-FOCK SELF-CONSISTENT FIELD THEORY

As we have seen, the electronic Hamiltonian contains two terms that act on one electron at a time, the kinetic energy and the electron-nucleus attraction, and a term that describes the pairwise repulsion of electrons. The latter depends on the coordinates of two electrons at the same time, and has turned out to be a practical computational bottleneck, which can be passed only for very small systems:

$$\hat{H}^1 = \sum_{i}^{n} \hat{H}_i^1 = -\frac{1}{2}\sum_{i=1}^{n} \Delta_i - \sum_{i=1}^{n}\sum_{A=1}^{N} \frac{Z_A}{|R_A - r_i|}$$

$$\hat{H}^2 = \sum_{i<j}^{n} \hat{H}_{ij}^2 = \sum_{i<j}^{n} \frac{1}{r_{ij}}$$

To avoid this problem the independent particle approximation is introduced: the interaction of each electron with all the others is treated in an average way. Suppose:

$$\hat{H}^2 = \sum_{i<j}^{n} \hat{H}_{ij}^2 = \sum_{i}^{n} \hat{V}_i^{av}$$

Then the Schrödinger equation which initially depended on the coordinates x (representing spatial and spin coordinates) of all electrons can be reduced to a set of equations:

$$\sum_{i=1}^{n}\left(\hat{H}_i^1+\hat{V}_i^{av}\right)\psi\left(x_1,x_2,\dots,x_n\right)=E\psi\left(x_1,x_2,\dots,x_n\right)$$

$$\left(\hat{H}_i^1+\hat{V}_i^{av}\right)\phi_i\left(x_1\right)=\hat{F}_i\phi_i\left(x_1\right)=\varepsilon_i\phi_i\left(x_1\right)$$

The wavefunctions $\phi_i\left(x_1\right)$ are called one-electron spin-orbitals.

The obvious problem is that for each electron the potential due to all other electrons has to be known, but initially none of these is known. In practice trial orbitals are used which are iteratively modified until a self-consistent solution (a "Self-Consistent Field") is obtained, which can be expressed as a solution to the Hartree-Fock equations:

$$\left(\hat{H}_i^1+\hat{V}_i^{av}\right)\phi_i=\hat{F}_i\phi_i=\varepsilon_i\phi_i$$

It is important to realise that *convergence of the SCF procedure is by no means guaranteed.* Many techniques have been developed over the years to speed up convergence, and to solve even difficult cases. In practice, difficulties often occur with systems with an unusual structure, where the electrons "do not know where to go".

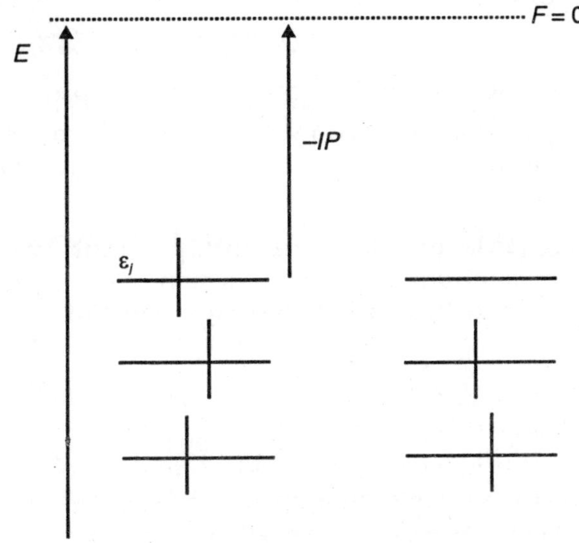

The eigenvalues are interpreted as orbital energies. The orbital energies have an attractively simple physical interpretation: they give the amount of energy necessary to take the electron out of the molecular orbital, which corresponds to the negative of the experimentally observable ionization potential (Koopmans' Theorem).

In addition to being a solution of the electronic Schrödinger equation the wavefunction must be normalized and satisfy the Pauli principle. The normalization condition is connected with the interpretation of the wavefunction as a distribution function which when integrated over entire space should give a value of one:

$$\int \psi' \psi \, dx = 1$$

in "bra-ket" notation:

$$\langle \psi | \psi \rangle = 1$$

The Pauli principle states that the wavefunction must change sign when two independent electronic coordinates are interchanged:

$$\psi_n(x_1, x_2, \ldots, x_i, \ldots, x_k, \ldots, x_m) = -\psi_n(x_1, x_2, \ldots, x_k, \ldots, x_i, \ldots, x_m)$$

For a two-electron system the spin-orbitals $\varphi_1(x_1, y_1, z_1, \sigma_1)$ and $\varphi_2(x_2, y_2, z_2, \sigma_2)$ (in which sigma is either alpha or beta spin state) can be combined as follows:

$$\psi_n = \frac{1}{\sqrt{2}} \{ \varphi_1(x_1, \ldots, \sigma_1) \varphi_2(x_2, \ldots, \sigma_2) - \varphi_1(x_2, \ldots, \sigma_2) \}$$

According to the definition of a determinant this antisymmetrized product is equal to:

$$\psi_n = \frac{1}{\sqrt{2}} \begin{vmatrix} \varphi_1(x_1, \ldots, \sigma_1) & \varphi_1(x_2, \ldots, \sigma_2) \\ \varphi_2(x_1, \ldots, \sigma_1) & \varphi_2(x_2, \ldots, \sigma_2) \end{vmatrix}$$

This type of wavefunction is known as a Slater determinant, commonly abbreviated as:

$$\psi_n = n^{-1/2}|\varphi_1\varphi_2...\varphi_n|$$

An important property of the SCF method is that its solutions satisfy the Variation Principle, which states that the expectation value of the energy evaluated with an inexact wavefunction is always higher than the exact energy:

$$E_\psi = \frac{\langle\psi|H|\psi\rangle}{\langle\psi|\psi\rangle} \geq E_{\psi exact}$$

As a consequence the lowest energy is associated with the best approximate wavefunction and energy minimization is equivalent with wavefunction optimization.

The energies of Slater determinants from a Hartree-Fock calculation are readily expressed in one- and two-electron integrals. For the ground state it is:

$$E = \sum_{i}^{OCC} H_{ii}^{i} + \sum_{i<j}^{OCC} \left[(ii\,|\,jj) - (ij\,|\,ij)\right]$$

Here we have used the following abbreviations:

$$H_{ii}^{1} = \left\langle i\left|\hat{H}_i^1\right|j\right\rangle = \int \varphi_i^*(x_i)\hat{H}_i^1\varphi_j(x_i)\,dx_i$$

$$(ij\,|\,kl) = \int\int \varphi_i^*(x_1)\varphi_j^*(x_1)\frac{1}{r_{12}}\varphi_k(x_2)\varphi_1(x_2)\,dx_1dx_2$$

The two-electron integral $(ii\,|\,jj)$ which describes the repulsion between two electrons each localized in one orbital is called a Coulomb integral, $(ij\,|\,ij)$ for which a classical picture cannot be drawn so easily is called the Exchange integral.

In many cases it is advantageous to apply the restriction that electrons with opposite spin pairwise occupy the same spatial orbital. This leads to the Restricted Hartree Fock method (RHF), as opposed to the Unrestricted version (UHF). An important advantage of the RHF method is that the magnetic moments associated with the electron spin cancel exactly for the pair of electrons in the same spatial orbital, so that the SCF wavefunction is an eigenfunction of the spin operators and . Note that the UHF wavefunction is more flexible than the RHF wavefunction, thus can approximate the exact solution better and give a lower energy. In practice RHF is mostly used for closed shell systems, UHF for open shell species. RHF models for open shell systems and more advanced models can used when necessary.

The total energy for a closed shell ground state RHF model can be written as:

$$E_{tot} = E_n + 2\sum_{i}^{occ} H_{ii} + \sum_{i,j}^{occ} (ii \mid jj) - \sum_{i,j}^{occ} (ij \mid ij)$$

The orbital energy in this case is:

$$\varepsilon_i = H_{ii} + \sum_{i,j}^{occ} \left[2(ii \mid jj) - (ij \mid ij) \right]$$

LIMITATIONS OF THE HF METHOD; ELECTRON CORRELATION

Restricted Hartree-Fock SCF theory has some painful shortcomings. Consider for example the dissociation of the H_2 molecule:

H^+ + H^- <——— H – H ————> H. + H.

A "dissociation catastrophe" occurs because the separated hydrogen atoms cannot be described using doubly occupied orbitals, so that H_2 tends to dissociate in H+ and H^-, which can be described with a doubly occupied orbital on H^-. This problem does not occur in the UHF method, but this method has the disadvantage that it does not give pure spin states.

An additional limitation of the HF method in general is that due to the use of the independent particle approximation the instantaneous correlation of the motions of electrons is neglected, even in the Hartree-Fock limit.

The difference between the exact energy (determined by the Hamiltonian) and the HF energy is known as the *correlation* energy: $E_{correlation} = E_{exact} - E_{HF} < 0$

Even though EHF is approximately 99% of Esub>exact the difference may be chemically important.

Several approaches are known that try to calculate the correlation energy after Hartree-Fock calculations (post-HF methods). We will very briefly discuss

- Configuration Interaction (CI),
- Moller-Plesset Perturbation Theory and
- Multi-Configuration SCF (MCSCF or CASSCF).

HF theory gives a wavefunction which is represented as a Slater determinant. In the conceptually simple Configuration Interaction (CI) method a linear combination of Slater determinants is constructed, using the unoccupied "virtual" orbitals from the SCF-calculation:

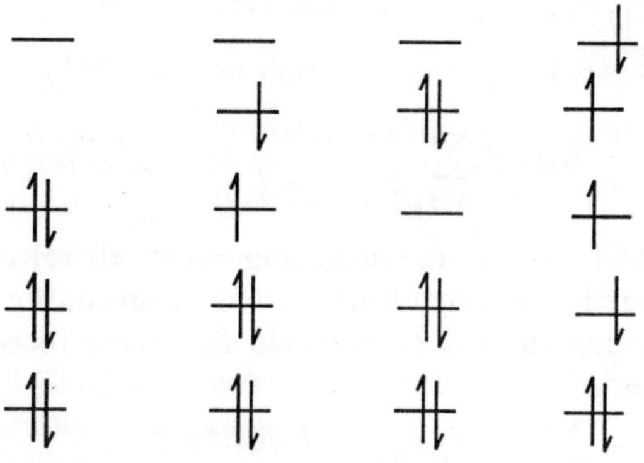

The total wavefunction is written as:

$$\psi = \phi_{HF} + \sum_{ij} C^{S}_{ij^*} \phi_i^{j^*} + \sum_{ijk^*l^*} C^{D}_{ijk^*l^*} \phi_{ij}^{k^*l^*} + \ldots$$

In principle, the exact correlation energy can be obtained from a full CI calculation in which all configurations are taken into consideration.

Unfortunately this is not possible for all but the smallest systems. Moreover, the problem is aggrevated when the size of the basis set is increased, on the way towards the Hartree-Fock limit. Thus, the theoretical limit of the exact (time-independent, non-relativistic) Schrödinger equation cannot be reached.

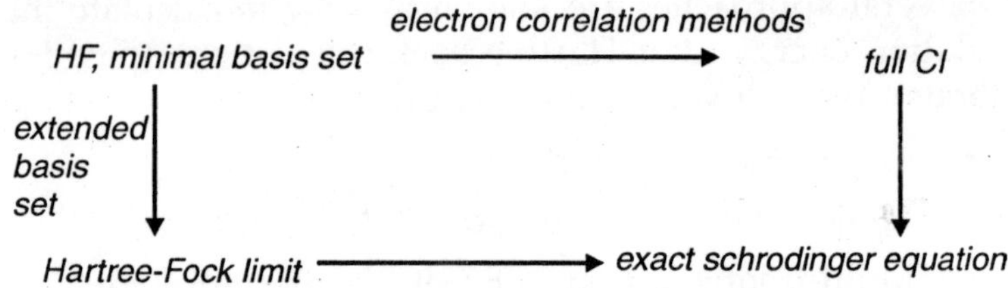

Even for small systems the number of excited configurations is enormously large. A popular way to truncate the CI expansion is to consider only singly and doubly excited configurations (CI-SD). The energy, calculated as the expectation value of the Hamiltonian for CISD is:

$$E = \frac{\langle \psi | H | \psi \rangle}{\langle \psi | \psi \rangle} =$$

$$E_{HF} + \sum_{i<j} \sum_{k^*<l^*} C^D_{ijk^*l^*} \left[\left(ij \mid k^*l^* \right) - \left(ik^* \mid jl^* \right) \right]$$

To perform the calculation one needs the two-electron integrals over Molecular Orbitals. The computation of these is very time-consuming, even when the integrals over AO's are available:

$$\left(ij \mid kl \right) = \sum_{\mu\nu\lambda\sigma} c_{i\mu} c_{j\nu} c_{k\lambda} c_{l\sigma} \left(\mu\nu \mid \lambda\sigma \right)$$

In general, CI is not the practical method of choice for the calculation of correlation energy because full CI is not possible, convergence of the CI expansion is slow, and the integral transformation time-consuming.

Moreover truncated CI is not size-consistent, which means that the calculation of two species at large separation does not give the same energy as the sum of the calculations on separate species. This is because a different selection of excited configurations is made in the two calculations. An advantage of the CI method is that it is variational, so the calculated energy is always greater than the exact energy.

Although CI is not recommendable as a method for ground states CI-singles (CIS) has been advocated as an approach to computation of excited state potential energy surfaces.

A different approach to electron correlation has become very popular in recent years: Møller-Plesset Perturbation Theory.

The basic idea is that the difference between the Fock operator and the exact Hamiltonian can be considered as a perturbation:

$$\hat{H} = \hat{F} + \hat{V}$$

Corrections can be made to any order of the energy and the wavefunction:

$$E = E_{HF} + E^{(1)} + E^{(2)} + E^{(3)} + \ldots$$

$$\psi = \psi_{HF} + \psi^{(1)} + \psi^{(2)} + \psi^{(3)} + \ldots$$

The most popular method is the lowest level of correction, MP2.

$$E_{corr}^{MP2} = \sum_{ijk^*l^*} \frac{2\left(ik^* \mid jl^*\right) - \left(ij \mid jk^*\right)}{\varepsilon_i + \varepsilon_j - \varepsilon_{k^*} - \varepsilon_l} \left(ik^* \mid jl^*\right)$$

An enormous practical advantage is that MP2 is fast (of the same order of magnitude as SCF), while it is rather reliable in its behaviour, and size consistent. A disadvantage is that it is not variational, so the estimate of the correlation energy can be too large. In practice MP2 must be used with a reasonable basis set (6-31G* or better). Subsequent MP-levels MP3, MP4 (usually MP4 SDQ) are more complicated and much more time-consuming.

For example, for pentane (C_5H_{12}) with the 6-31G(d) basis set (99 basis functions) an MP2 energy calculation took about 4 times the amount of time needed for SCF, while MP4 took almost 90 times that time.

Multiconfiguration SCF (MCSCF) or Complete Active Space SCF (CASSCF) is a special method in which HF-orbitals are optimized simultaneously with a "small" CI.

This can be used to study problems where the Hartree-Fock method is inappropriate (e.g. when there are low-lying excited states), or to generate a good starting wavefunction for a subsequent CI calculation.

$$\psi_{MCSCF} = \sum_{k=1}^{n} C_k \left|\phi_{k_1}\phi_{k_2}\phi_{k_3}\cdots\phi_{k_n}\right|$$

The MCSCF method requires considerable care in the selection of the basis set and especially the active space, and should not be considered for routine use.

In contrast to the HF, MPn and CI methods, MCSCF does not provide a "model chemistry" because each problem requires different choices.

MCSCF methods are essential for the study of processes in which transitions between potential energy surfaces occur, such as in photochemical reactions.

Other methods to determine the correlation energy are under development.

At this point it is useful to note another promising development, that of *density functional theory*. This is a method in which the two-electron integrals are not computed in the conventional way. Application of this approach to molecular systems is still in its infancy, but rapid developments are to be expected in the next few years, in particular driven by the desire to be able to compute larger systems, e.g. metal complexes and organometallic compounds.

ENERGY CALCULATIONS

Ab initio calculations give the absolute energy of the system of fixed nuclei and moving electrons. These are large numbers, for example for cyclohexane the HF energy with the 6-31G* basis set is –234.2080071 a.u., which is equal to 146967.86 kcal/mol.

Thus, the chemically significant energy quantities of a few kcal/mol are very much smaller than the computed quantity, and high accuracy is required.

The absolute energy is not a directly useful quantity. It can however be used to calculate the Heat of Formation with a reasonable accuracy. According to G1 and G2 theories the molecular structure and vibrational frequencies are first determined at the HF/6-31G* level.

The frequencies are used to calculate the zero-point energy. Then, the geometry is further optimized at the MP2 level. Subsequently, basis set effects and correlation energies are calculated at various levels of theory, to allow an extrapolation (using small empirical contributions !) to the limits of full CI and the Hartree-Fock limit, that is to the complete Schrödinger equation for the motionless molecule.

Finally, the zero-point vibrational energy is added. This procedure can account for heats of formation with an accuracy of < 2 kcal/mol, which rivals the quality of experimental data.

This is a much less elaborate procedure than the G1 and G2 theories, but it is essentially empirical. The empirical corrections needed in G_1 and G_2 are of a very "mild" kind, they are not related to the structure of the species, but only depend on the number of electrons.

The *isodesmic reaction* approach allows a fairly accurate calculation of the heat of reactions, even at the HF level. Isodesmic reactions are defined as transformations in which the numbers of bonds of each formal type are conserved, and only the relationships among the bonds are altered. For example:

$$CH_4 + CH_3CH_2OH \rightarrow CH_3CH_3 + CH_3OH \tag{1}$$

$$CF_4 + 3CH_4 \rightarrow 4CH_3F \tag{2}$$

Energy changes (kcal/mol) for these two reactions are:

	STO-3G	3-21G	6-31G*//STO-3G	experimental
deltaE (1)	2.6	4.8	4.1	5.0 (5.7)
deltaE (2)	53.5	62.4	49.6	49.3 (52.8)

(The experimental numbers in parentheses are without correction for zero-point energy changes).

QUALITY OF AB INITIO RESULTS

One of the most useful features of ab initio MO theory is that it allows the definition of "model chemistries". A theoretical model chemistry entails a method (e.g. Hartree-Fock or MP2 etc.) and a basis set.

The philosophy of a model chemistry is that it should be uniformly applicable and tested on as many systems as possible to learn about its performance. This turns out to be useful, because the reliability and accuracy of model chemistries can be systematically assessed in this way.

In the rest of this section we will focus on the performance of model chemistries that can be practically applied for organic molecules with present-day hardware and software, that is HF and MP2 methods with basis sets usually limited to the 6-31G(d) level.

Geometry

As far as equilibrium geometry is concerned, HF and MP2 ab initio models even with modest basis sets lead to excellent results. At present, HF/6-31G* or MP2/6-31G* are considered good and reliable methods for the determination of the geometries of organic molecules. In many cases the smaller basis sets 3-21G or even STO-3G can give useful results. The bond length calculated at the HF level is usually overestimated by ca. 0.01 - 0.02 Å as a result of the neglect of electron correlation.

For transition metal compounds and organometallics the results are less satisfactory. Because of the size of such systems, adequately large basis sets cannot be applied with the present generation of computers and programmes. Moreover, electron correlation can be important. It is conceivable that a model chemistry based on Density Functional Theory will become available which covers this area of chemistry.

Vibrational Frequencies

Due to the availability of analytic second derivatives of HF and MP2 wavefunctions, the calculation of vibrational frequencies and normal modes of organic molecules has become almost a routine matter. It turns out that the results even with modest-basis HF models are quite good.

The frequencies are consistently overestimated, which is due to the neglect of correlation energy and of anharmonicity. Uniform scaling of the computed frequencies by a factor of 0.89 ± 0.01 gives a good agreement for most cases. For MP2 the scaling factor should be closer to 1.0.

Of course, with dedicated empirical force fields a better fit of the spectra can be achieved, however at the expense of generality and of great effort.

Energies

The accurate computation of absolute or relative energies remains a major challenge. Even conformational energy differences and barriers are not reliably computed with HF or MP2 models using small basis sets (6-31G* or smaller).

Of course the demands on accuracy are very high in this case. On the other hand, a comparison between related systems (e.g. predicting a substituent effect) can often be made quite well.

Energies of reactions can be predicted relatively accurately, especially for isodesmic processes. When the number of formal bonds changes, electron correlation methods are essential. As mentioned successful methods have been developed to estimate heats of formation on the basis of ab initio results.

Chemical Reactions

The quality of the prediction of structures of transition states can hardly be verified by comparison with experiments, so the only way is to look at convergence of the computed values with increasing sophistication of the method employed.

Energies of transition states can be related to experimental activation energies.

In practice theoretical methods can at best to predict relative activation energies, in other words the selectivities of reactions. Often this is chemically more significant than the precise number: usually it doesn't really matter whether a product ratio is 50: 1 or 100: 1. To be able to predict whether products will be formed in a ratio of 3:1 or 1:3 one needs relative TS energies with an accuracy of roughly 1 to 2 kcal/mol, and this is feasible for some reactions, such as the Diels-Alder reaction.

SEMI-EMPIRICAL QUANTUM CHEMISTRY

Ab initio quantum chemical methods are limited in their practical applicability because of their heavy demands of cpu-time and storage space on disk or in the computer memory.

At the Hartree-Fock level the problem is seen to be in the large number of two-electron integrals that need to be evaluated. Without special tricks this is proportional to the fourth power of the number of basis functions. In practice this can be reduced to something close to the third power for larger molecules, e.g. because use is made of the fact that integrals between orbitals centered on distant atoms need not be calculated because they will be zero anyway.

Still, the size of systems that can be treated is limited, and this holds much more strongly for correlated treatments.

MP2 for example formally scales with the fifth power of the number of basis functions. Therefore there is a place for more approximate methods that retain characteristics of the quantum-chemical approach, in particular the calculation of a wavefunction from which electronic properties can be derived. In this section we will present a brief overview of commonly used semi-empirical methods.

The semi-empirical methods are based on the Hartree-Fock approach. A Fock-matrix is constructed and the Hartree-Fock equations are iteratively solved. The approximations are in the construction of the Fock matrix, in other words in the energy expressions. Recall how the Fock matrix elements are expressed as integrals over atomic basis functions:

$$F_{\mu\nu} = \langle \mu | \hat{F} | \nu \rangle = \langle \mu | \hat{H}_1 | \nu \rangle + \sum_{\lambda,\sigma} P_{\lambda\sigma} \left[(\mu\nu | \lambda\sigma) - \frac{1}{2} (\mu\lambda | \nu\sigma) \right]$$

in which P is the density matrix:

$$P_{\lambda\sigma} = 2 \sum_{i}^{occ} c_{i\lambda} c_{i\sigma}$$

To simplify matters drastically, the Zero Differential Overlap (ZDO) approximation assumes:

$$\varphi_\mu(r)\varphi_\nu(r) = 0 \quad \text{for } \mu \neq \nu$$

which implies that

$S_{\mu\nu} = \langle \mu | \nu \rangle = \delta_{\mu\nu}$ in which $S_{\mu\nu} = 0$ if $\mu \neq \nu, \delta_{\mu\nu} = 1$

This can be justified when the atomic basis orbitals are orthogonalized (Löwdin orthogonalization).

As a result of the ZDO approximation many two-electron integrals vanish:

$$(\mu\nu | \lambda\sigma) = \delta_{\mu\nu}\delta_{\lambda\sigma} (\mu\mu / \lambda\lambda)$$

Another common feature of semi-empirical methods is that they only consider the valence electrons.

The core electrons are accounted for in a core-core repulsion function, together with the nuclear repulsion energy.

In the most popular semi-empirical methods used today (MNDO, AM1 and PM3) the ZDO approximation is only applied to basis functions on different atoms. This is called the NDDO approximation.

The next step is to replace many of the remaining integrals by parameters, which can either have fixed values, or depend on the distance between the atoms on which the basis functions are located. At this stage empirical parameters can be introduced, which can be derived from measured properties of atoms or diatomic molcules.

In the modern semi-empirical methods the parameters are however mostly devoid of this physical significance: they are just optimized to give the best fit of the computed molecular properties to experimental data.

Different semi-empirical methods differ in the details of the approximations (e.g. the core-core repulsion functions) and in particular in the values of the parameters. Note that in contrast to molecular mechanics, only parameters for single atoms and for atom pairs are needed. The number of published parameters increases steadily.

The semi-empirical methods can be optimized for different purposes. The MNDO, AM1 and PM3 methods were designed to reproduce heats of formation and structures of a large number of organic molecules. Other semi-empirical methods are specifically optimized for spectroscopy, e.g. INDO/S or CNDO/S, which involve CI calculations and are quite good at prediction of electronic transitions in the UV/VIS spectral region.

Some even more approximate methods are still quite useful. In the Hückel and Extended Hückel methods the whole sum over two-electron integrals is replaced by a single diatomic parameter (the resonance integral), so that no search for a self-consistent field is necessary (nor possible). These methods have proven extremely valuable in qualitative and semi-quantitative MO theories of pi-electron systems and of organometallic systems.

For pi-electron systems ZDO treatments have been developed that take only pi-centers (p-atomic orbitals) into account, but do perform the SCF calculation. An example is the Pariser-Parr-Pople method, which involves a CI calculation as well. This method is very successfully used to predict the optical absorption spectra of conjugated organic molecules.

In the MM2 and MM3 programmes pi-electron calculations are used to adjust the force constants and equilibrium values of bond lengths to the prevailing bond order. The pi-bond order between two atoms is simply the sum over MOs of the product of the coefficients of the basis functions on the atoms in the MO, multiplied by the occupation number of the MO:

$$P_{rs}^{\pi} = \sum_{i} n_i c_{ri} c_{si}$$

For a given geometry the pi-electron calculation is done, and the bond-orders computed. Then the force field is adjusted: the force constants for stretching and torsion are scaled and the equilibrium bond length for the bonds between the pi-centers are calculated.

$$k_{s_{ij}} = k_s(0) + c_s p_{ij}^{\pi}$$

$$r_{0_{ij}} = r_0(0) + d_s p_{ij}^{\pi}$$

When the geometry changes too much, the pi-electron treatment is repeated to adjust the force field to the new situation. For the pi-electron calculation the pi-system is treated as if it is planar. Otherwise the bond order for a twisted semi-single bond would become smaller as the bond is twisted more, and the "restoring force" towards planarity (conjugation) would vanish.

QUALITY OF SEMI-EMPIRICAL RESULTS

Semi-empirical methods are parameterized on the basis of selected properties of a selected set of molecules. A reasonable performance can be expected for related compounds, but for problems that are not covered by the "training set" the reliability is limited. In this respect ab initio methods, even low-level ones, are more widely applicable.

For molecular structure and heats of formation of closed-shell molecules MNDO, AM1 and PM3 are quite good. Practical experience has shown that for some particular problems one of the three performs markedly better than the others, but in general the most recent methods AM1 and PM3 are preferred. PM3 is parameterized for a greater number of elements, but sometimes the parameters are based upon a very small set of data. The mean absolute errors of the bond lengths between heavy atoms are reported to be 0.036 Å for PM3 and slightly greater for AM1 and MNDO. The error in bond angles is 3 - 4 degrees. Even low-level ab initio calculations are usually better, but very much more time-consuming.

Trends in vibrational frequencies in related systems are usually reproduced by semi-empirical calculations, but the errors are not as nicely systematic (and therefore correctable) as in ab initio calculations. When energy is concerned the results are not completely satisfactory, in spite of the fact that the methods are parameterized to reproduce heats of formation.

SOLVATION

The effect of a solvent can be incorporated in quantum-chemical calculations most easily by considering it as a continuous dielectric medium, characterized by a dielectric constant. The electric field caused by the molecule induces a polarization of the medium, which in turn acts on the electrons in the molecule (Self-Consistent Reaction Field, SCRF).

The model thus contains the quantum-mechanical description of the molecule and a classical medium. The problem is to choose where to locate the boundary between quantum system and medium. In the Gaussian programmes a simple approximation is used in which the volume of the solute is used to compute the radius of a cavity which forms the hypothetical surface of the molecule. Spartan offers solvation models for the semi-empirical Hamiltonians.

This method, in which the molecular surface is constructed from atomic (Born) radii, requires parameters for each atom. In most cases, solvation hardly affects the structure of a molecule (relative to the gas phase), but in cases of polar molecules, zwitterions or ions, the relative energies can be changed dramatically.

PROPERTIES DERIVED FROM THE WAVEFUNCTION

The electronic wavefunction which is computed in ab initio as well as semi-empirical quantum chemical methods can be used to derive observable quantities of a molecule, but it can also be analysed and used to rationalize certain chemical phenomena.

Electrical Properties

The electric dipole moment μ of a molecule can be calculated directly from the positions of the nuclei and the electronic wavefunction:

$$\mu = 2.5416\left(\sum_{A}^{nuclei} Z_A r_A - \sum_{\mu}^{N}\sum_{v}^{N} P_{\mu v}\langle\mu|r|v\rangle\right)$$

The dipole moment can be viewed as the first term of an expansion of the electric field due to the molecule, the next higher term being the quadrupole moment. It is also possible to obtain the dipole moment and polarizabilities directly as derivatives of the energy with respect to a uniform electric field. The electrostatic potential of the molecule represents the interaction between the charge distribution of the molecule and a unit point charge located at some position p:

$$\varepsilon_p = \sum_{A}^{nuclei} \frac{Z_A}{R_{Ap}} - \sum_{\mu}^{N}\sum_{v}^{N} P_{\mu v} \frac{\langle\mu|v\rangle}{r_{1p}}$$

Calculation of the molecular electrostatic potential at the surface of the molecule (described by the total electron density) can indicate how the molecule will interact with polar molecules or charged species. Visualization of this can be nicely accomplished using colour coding.

Atomic Charges

Although concepts like atomic point charges or bond dipoles are widely used in molecular mechanics, there is no unique definition of atomic charge in a molecule. All ways to attribute a part of the electron density to individual atoms are to a certain extent arbitrary. As a first analysis, or as a way to compare related systems, Mulliken Population Analysis can be applied. The electron density distribution (the probability of finding an electron in a volume element dr) is:

$$\rho(r) = \sum_{\mu}^{N}\sum_{V}^{N} P_{\mu v}\varphi_{\mu}\varphi_{v}$$

Integrated over entire space this gives the total number of electrons (Sμ*v* is the overlap):

$$\int \rho(r)\,dr \quad = \sum_{\mu}^{N}\sum_{v}^{N} P_{\mu v}S_{\mu v} = n$$

This can be separated into diagonal and off-diagonal terms, where the former represent the net population of the basis orbitals and the latter are make up the overlap population.

$$\sum_{\mu}^{N} P_{\mu v} + 2\sum_{\mu}^{N}\sum_{<v}^{N} P_{\mu v}S_{\mu v} = n$$

$P_{\mu\mu}$ net population of φ_{μ}

$Q_{\mu v} = 2P_{\mu v}S_{\mu v}$ =overlap population of φ_{μ} and φ_{v}

In the Mulliken scheme the overlap population is simply shared between the contributing atoms, which leads to the following charge for each basis orbital:

$$q_{\mu} = P_{\mu\mu} + \sum_{\mu \neq v} P_{\mu v}S_{\mu v}$$

Summing of the charges in the orbitals associated with each atom gives the atomic charge.

An important disadvantage of the Mulliken population analysis is that extended basis sets can lead to unphysical results, e.g. charges of more than 2e, which result from the fact that the basis orbitals centered at one atom actually describe electron density close to another nucleus. Population

Analysis based on *Natural Atomic Orbitals* does not have this problem.

An approach which may be physically more relevant is to fit charges at the atomic positions to the *molecular electrostatic potential* measured at a grid of points. This still leaves some arbitrariness in the choice of the grid, and the procedure is computationally much more demanding than the other types of population analysis.

Effect of Molecular Symmetry on Melting Temperature and Solubility

Molecular symmetry has a pronounced effect on the melting properties and solubility of organic compounds. As a general rule, symmetrical molecules in crystalline form have higher melting temperatures and exhibit lower solubilities compared with molecules of similar structure but with lower symmetry. Symmetry in a molecule imparts a positive amount of residual entropy in the solid phase (*i.e.*, more possible arrangements leading to the same structure). This means that the entropy of a crystal of symmetric molecules is greater than the entropy of crystal of a similar, but non-symmetric molecule. An analysis is presented relating the enthalpy, entropy and temperature of melting for an idealised system of structural isomers of different molecular symmetries.

The analysis presented helps explain why often, yet not always, the crystal of a more symmetric molecule, which has greater entropy to start (closer to that of the liquid), also exhibits a greater gain in entropy upon melting, compared with the crystal of a less symmetrical molecule. The residual entropy due to molecular symmetry has the direct effect of reducing the entropy gain upon melting (a negative effect). However, molecular symmetry also exerts indirect effects on both the entropy and enthalpy of melting. These indirect effects, imposed by the condition of equilibrium melting, are positive, such that it is the balance between the direct and indirect effects what determines the value observed for the entropy of melting of the symmetric molecules. When the indirect effect of molecular symmetry is greater than its direct effect, the observed entropy gain upon melting of the more symmetrical molecule is greater than that of a less symmetrical one.

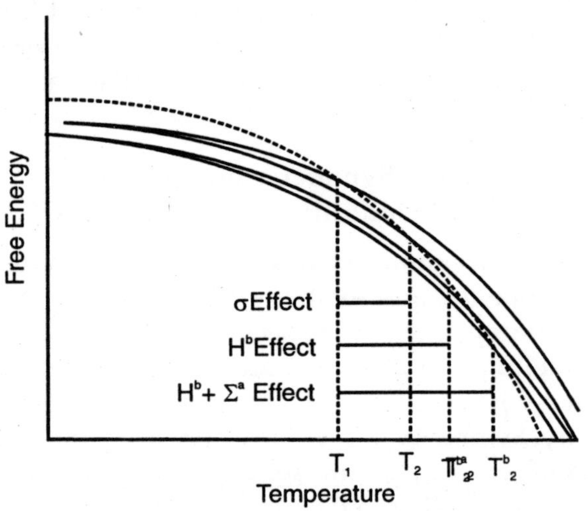

THE STANDARD MODEL

The most fundamental known particles of the universe, and how they act as building blocks for everything that we know. The theory describing this scheme is called the standard model. Speculations exist about possible more fundamental structures in the universe, such as the constructs of string theory. However, with the standard model we have reached the frontier of what is known with any degree of certainty.

QUARKS AND LEPTONS

The standard model of quarks and leptons is a united set of quantum mechanical theories encompassing electromagnetism, the weak force, which is responsible for beta decay, and the strong force, which holds atomic nuclei together. Before investigating the standard model, we need to describe the state of affairs previous to its development. The creation of high energy particle accelerators led to the discovery of a plethora of particles in addition to those already known. These particles fall into the following categories:

- *Leptons* are spin 1/2 particles which do not interact via the strong force. The electron, muon, and the electron and muon neutrinos are examples.
- *Hadrons* are particles which interact via the strong force. They are divided into two sub-categories depending on their spin:
  - *Baryons* are hadrons with half-integral spin, mainly 1/2 and 3/2. The proton and neutron are well known examples. The neutral lambda particle is another.
  - *Mesons* are hadrons with integral spin, mainly 0 and 1. Examples are the pions and kaons.
- *Strange particles* are baryons and mesons which are unstable, but have much longer half-lives than other particles of similar mass and spin. This is interpreted to mean that such particles possess a property called *strangeness* which is conserved by strong processes, thus making strange particles stable against strong decay into non-strange particles. However, strangeness is not conserved by weak processes, allowing strange particles to decay via the weak interaction. This explains their anomalously long half-lives. Strange particles are always created in pairs by strong processes in such a way that the total strangeness remains zero. For instance, if one particle has strangeness +1 then the other must have strangeness –1. An example of strange particle production is when a negative pion collides with proton, giving rise to a neutral lambda particle and a neutral kaon.
- *Intermediary particles* are those which transfer energy, momentum, charge, and other properties from one particle to another in association with one of the four fundamental forces.

- *Photons* transmit the electromagnetic force and have zero mass and spin 1.
- *Gravitons* are thought to transmit the gravitational force, though they have not been directly observed. The graviton is postulated to have zero mass and spin 2.

We will discover additional intermediary particles in our discussion of the standard model.

- *Antiparticles* exist for all particles. These have the same mass and spin but opposite values of the electric charge and various other quantum numbers such as lepton number or baryon number. The lepton number is the number of leptons minus the number of antileptons, with a similar definition for baryon number. Thus, a lepton has lepton number 1 and a baryon has baryon number 1. Their antiparticles have lepton number –1 and baryon number –1. As far as we know, baryon number and lepton number are absolutely conserved, which means that baryons and leptons can only be created or destroyed in particle-antiparticle pairs. Antiparticles are represented by the symbol of the particle with an overbar.

QUANTUM CHROMODYNAMICS

The standard model postulates that all known particles are either fundamental point particles or are composed of fundamental point particles according to a remarkably small set of rules. Just as atoms are bound states of atomic nuclei and electrons, atomic nuclei are bound states of protons and neutrons. Atomic nuclei are discussed in the next section. In this section we delve one step deeper in the heirarchy of the universe. We now believe that all hadrons are actually bound states of fundamental spin 1/2 particles called quarks. Whereas all other known particles have an electric charge equal to where is the proton charge, quarks have electric charges equal to $\pm e$ either $-e/3$ or $+2e/3$. Leptons themselves are considered to be fundamental, so the leptons and the quarks form the basic building blocks of all matter in the universe.

Quarks are subject to electromagnetic forces via their charge, but interact most strongly via the so-called strong force. The strong force is carried by massless, uncharged, spin 1 bosons called gluons.

Table: Table of quark types, charge (as a fraction of the proton charge), rest energy (in GeV), and the four "exotic" flavour quantum numbers.

Type	Charge	Rest energy	s	c	b	t
down (d)	-1/3	0.333	0	0	0	0
up (u)	+2/3	0.330	0	0	0	0
strange (s)	-1/3	0.486	-1	0	0	0
charm (c)	+2/3	1.65	0	+1	0	0
bottom (b)	-1/3	4.5	0	0	-1	0
top (t)	+2/3	176	0	0	0	+1

When Murray Gell-Mann and George Zweig first proposed the quark model in 1963, they needed to postulate only three types or flavours of quarks, up, down, and strange. These were sufficient to explain the constitution of all hadrons known at the time. We currently know of six different flavours of quarks. Their properties are listed in table. The properties charm, topness, and bottomness are analogous to strangeness these properties are conserved in strong interactions. Weak interactions, discussed in the next section, can turn quarks of one flavour into another flavour. However, the strong and electromagnetic forces cannot do this.

Table: Sample hadrons. The charge is specified as a fraction of the proton charge, the rest energy is in MeV, and the mean life (1.44 times the half life) is in seconds. The composition is presented in terms of the flavours of quarks and antiquarks which make up the particle.

Type	Charge	Rest energy	Spin	Composition	Mean life
Proton	+1	938.280	1/2	uud	stable
Neutron	0	939.573	1/2	udd	898
Lambda	0	1116	1/2	uds	3.8×10^{-9}
positive pion	+1	140	0	$u\,\overline{d}$	2.6×10^{-8}
negative pion	-1	140	0	$\overline{u}\,d$	2.6×10^{-8}

neutral pion	0	135	0	u $\bar{u}$ - d $\bar{d}$	8.7×10^{-17}
positive rho	+1	770	1	u $\bar{d}$	4×10^{-24}
positive kaon	+1	494	0	u $\bar{s}$	1.24×10^{-8}
neutral kaon	0	498	0	d $\bar{s}$	8.6×10^{-11}
J/psi	0	3097	1	c $\bar{c}$	1.5×10^{-20}

Just as the proton and the neutron have antiparticles, so do quarks. Antiquarks of a particular type have strong and electromagnetic charges of the sign opposite to the corresponding quarks. Quarks have baryon number equal to 1/3, while antiquarks have –1/3. Thus combining three quarks results in a baryon number equal to 1, while together a quark plus an antiquark have baryon number zero. All baryons are thus combinations of three quarks, while all mesons are combinations of a quark and an antiquark. Table lists a sampling of hadrons and some of their properties. Notice that the same combination of quarks can make up more than one particle, e. g., the positive pion and the positive rho. The positive rho may be considered as an excited state of the u $\bar{d}$ system, while the positive pion is the ground state of this system.

Yet to be mentioned is the quantum number colour, which has nothing to do with real colours, but has analogous properties. Each flavour of quark can take on three possible colour values, conventionally called red, green, and blue. This is illustrated in figure. Antiquarks can be thought of as having the colours antired, antigreen, and antiblue, also known as cyan, magenta, and yellow. Because of this, the theory of quarks and gluons is called quantum chromodynamics. Counting all colour and flavour combinations, there are $6\times3=18$ known varieties of quarks.

As in electromagnetism, the strong force has associated with it a "strong charge", g_s. However, this charge is somewhat more complicated than electromagnetic charge, in that there are three kinds of strong charge, one for each of the strong force colours. Each colour of charge can take on positive and negative values equal to $\pm g_S$. As with electromagnetism, positive and negative charges (of the same colour) cancel each other. However, in quantum chromodynamics there is an additional way in which charges can cancel. A combination of equal amounts of red, green, and blue charges results in zero net strong charge as well.

Gluons, the intermediary particles of the strong interaction come in eight different varieties, associated with differing colour-anticolour combinations. Since gluons don't interact via the weak force, there is no flavour quantum number for gluons — quarks of all flavours interact equally with all gluons.

The quark model of matter has led to extensive searches for free quark particles. However, these searches for free quarks have proven unsuccessful. The current interpretation of this result is that quarks cannot exist in a free state, basically because the attractive potential energy between quarks increases linearly with separation. This appears to be related to the fact that gluons, the intermediary particles for the strong force, can interact with each other as well as with quarks. This leads to a series of increasingly complex processes as quarks move farther and farther apart. The result is called quark confinement — apparently, individual quarks can never be observed outside of the confines of the observable particles which contain them.

Confinement works not only on single quarks, but on any "coloured" combinations of quarks and gluons, e. g., a red up quark combined with a green down quark. It appears that long range inter-quark forces only vanish for interactions between "white" or "colour-neutral" combinations of quarks. This is why only colour-neutral combinations of quarks — three quarks of three different colours or a quark-antiquark pair of the same colour — are actually seen as observable particles.

The strong equivalent of the fine structure constant is the coupling constant for the strong force:

$$\alpha_s = \frac{g_s^2}{4\pi\hbar c}$$

Note that α_s is dimensionless. The binding energy between quarks is comparable to the rest energies of the quarks themselves. In other words, $\alpha_s \approx 1$. Furthermore, as we have noted, the potential energy between two quarks appears to increase indefinitely with separation. Though forces exist between colour-neutral particles, they are weak and of short range compared to the forces between quarks or coloured combinations of quarks.

However, they are still relatively strong compared to, say, electromagnetic forces. As we shall see later, these residual strong forces are responsible for nuclear processes.

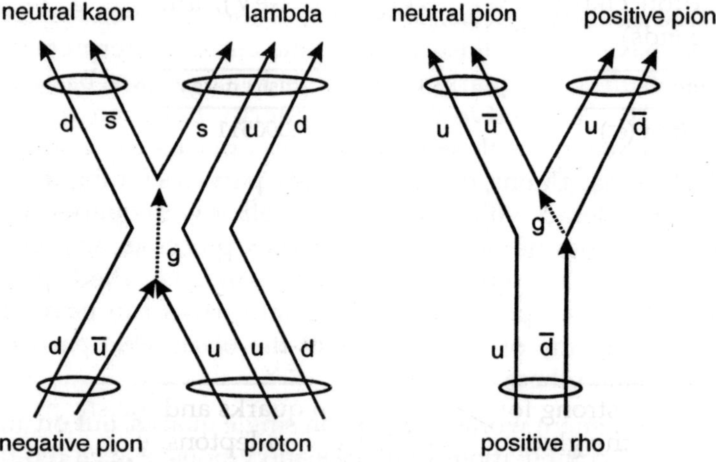

Fig. Some sample strong interactions illustrated in terms of gluon emission and absorption. The process on the left shows the reaction $\pi^- + p \rightarrow K^0 + \Lambda^0$, while the one on the right shows the decay $\rho^+ \rightarrow \pi^+ + \pi^0$. Quarks are labeled with solid lines while gluons are shown by dashed lines.

Interactions between hadrons can be thought of as resulting from interactions between the individual quarks making up the hadrons. Two sample strong interactions are shown in figure. Virtual gluons can be emitted and absorbed by quarks much as virtual photons can be emitted and absorbed by electrically charged particles. Particles unstable to strong decay processes (such as the positive rho particle) typically live only about $10^{-23}\,s$, where as particles stable to strong decay but unstable to weak decay live of order $10^{-10}s$ or longer, depending strongly on how much energy is liberated in the decay. Particles subject to electromagnetic decay processes, such as the neutral pion, take on mean lifetimes intermediate between strong and weak values, typically of order $10^{-18}s$.

THE ELECTROWEAK THEORY

Table: Table of lepton types, charge (as a fraction of the proton charge), rest energy (in GeV), and mean life (in seconds).

Type	Charge	Rest energy	Mean life
electron (e^-)	-1	0.000511	stable
electron neutrino (v_e)	0	≈ 0	stable
muon (μ^-)	-1	0.106	2.2×10^{-6}
mu neutrino (ν_μ)	0	≈ 0	stable
tau (τ)	-1	≈ 1.7	3.0×10^{-13}
tau neutrino (ν_τ)	0	≈ 0	stable

The strong force acts only on quarks and the strong force carrier, the gluon. It does not act on leptons, e. g., electrons, muons, or neutrinos. Table shows all of the known leptons. The so-called weak force acts on leptons as well as on quarks.

In 1979 Sheldon Glashow, Abdus Salam, and Steven Weinberg won the Nobel Prize for their electroweak theory, which unites the electromagnetic and weak interactions. Unlike the strong and electromagnetic forces, the intermediary particles of the weak interaction, the W^+, the W^-, and the Z^0, have rather large masses. In particular, the rest energy of the $W^{\pm}$ is $81GeV$ while that of the Z^0 is 92GeV. Electroweak theory considers electromagnetism and the weak interactions to be different aspects of the same force. A key aspect of the theory is the explanation of why three out of four of the intermediary particles of the electroweak force are massive. (The photon is the massless one.) Unfortunately, the details of why this is so are highly technical, so we cannot delve into this subject here. We only note that the explanation requires the existence of a highly massive (several thousand GeV) spin zero boson called the Higgs particle. Due to its large mass, we have not yet determined whether the Higgs particle exists.

The weak force has certain bizarre properties not shared by the other forces of nature:

- The weak interaction can change quark flavours. For instance, the beta decay of a neutron converts a down quark into an up quark. On the other hand, the weak interaction is "colourblind", i. e., it is insensitive to quark colours.
- The weak interaction is not left-right symmetric. In other words, the physical laws governing the weak interaction look different when seen in a mirror.
- The weak interaction is slightly asymmetric to the interchange of particles and antiparticles in certain situations.

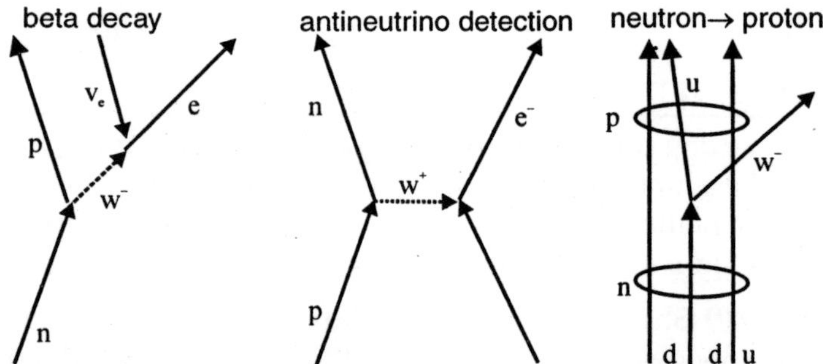

Fig: Illustration of two weak reactions. The left panel shows beta decay while the middle panel shows how electron antineutrinos can be detected by conversion to a positron. The right panel shows how W^- emission works according to the quark model, resulting in the conversion of a down quark to an up quark and the resulting transformation of a neutron into a proton.

The prototypical weak interaction is the decay of the neutron into a proton, an electron, and an antineutrino. This decay is energetically possible because the neutron is slightly more massive than the proton, and is illustrated in the left panel of figure. Note that this figure is drawn as if a neutrino moving backward in time absorbs a W^- particle, with a resulting electron exiting the reaction forward in time. However, we know that this is equivalent to an electron and an antineutrino both exiting the reaction forward in time according to the Feynman interpretation of negative energy states.

The weak interaction is called "weak" because it appears to be so in commonly observed processes. For instance, the range of a relativistic electron in ordinary matter is of order centimeters to meters. This is because the electromagnetic force between the charge of the electron and the charges on atomic nuclei are strong enough to rapidly cause the energy of the electron to be dissipated. However, the range in matter of a neutrino produced by beta decay is many orders of magnitude greater than that of an electron. This is not because the weak force is intrinsically weak — the value of the "fine structure constant" for the weak force is $\alpha_w \approx 10^{-2}$ according to the standard model, and is actually larger than α for electromagnetism.

The real reason for the apparent weakness of the weak force is the large mass of the intermediary particles. As we have seen, large mass translates into short range for a virtual particle at low momentum transfers. This short range is what causes the weak force to appear weak for momentum transfers much less than the masses of the *W* and *Z* particles, i. e., for $q \ll 100GeV$. For leptons and quarks with energies $E \gg 100GeV$, the weak force acts with much the same strength as the electromagnetic force.

GRAND UNIFICATION?

Table: Generations of leptons and quarks. Members of each generation tend to fit together.

Generation	Leptons	Quarks
1	electron	down
	electron neutrino	up
2	muon	strange

	mu neutrino	charm
3	tau	bottom
	tau neutrino	top

The standard model is a great achievement, but it leaves a number of questions unanswered. As table shows, nature seems to have produced more particles than are needed to construct the universe. Virtually everything we know of is composed of electrons, electron neutrinos, up quarks, and down quarks. These four particles seem to fall naturally together in a family or generation. Why then are there apparently unneeded additional generations? What role do muons, taus, and the exotic quark forms play in the universe?

Another question concerns the dichotomy between leptons and quarks. Electrons and electron neutrinos can be converted into each other by weak interactions, as can up and down quarks.

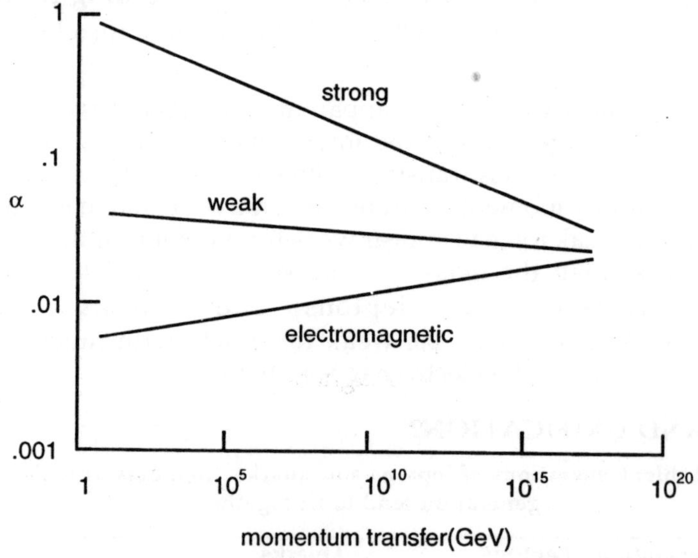

Fig: Speculated behaviour of the dependence of the coupling constant α on momentum transfer for each of the forces. Extrapolation based on current measurements suggests that these constants come together to a common value at very high momentum transfer.

In the standard model, electromagnetic and weak forces are truly united as aspects of a single phenomenon. However, quantum chromodynamics stands more on its own. One could imagine further advances which would show that the electroweak and strong forces were in fact different aspects of the same phenomenon. This could be characterized as a grand unification of the forces of nature.

As previously noted, the strong force coupling constant, α_s, gets smaller with increasing momentum transfer. It turns out that the weak coupling constant, α_W, exhibits similar behaviour, while the electromagnetic coupling constant, the fine structure constant α, becomes stronger at higher energies. This behaviour is illustrated in figure, though it is based on data only up to about $10^3 GeV/c$. Figure is thus largely speculative. However, if the observed trends do continue to very high momentum transfers, this would be evidence in favour of grand unification.

A number of speculative grand unification theories have been proposed. Most such theories view leptons and quarks as being different states of the same particle and also predict that leptons can turn into quarks and vice versa, albeit at very low rates. One of the consequences of such theories is that the proton would be an unstable particle, but with a very long lifetime, of order 10^{30}yr. Experiments have been done to detect the decay of the proton, but so far without success. These experiments are sufficient to rule out some but not all of the proposed grand unification theories.

One task which would not be accomplished by grand unification is the incorporation of gravity into a common framework with the strong, weak and electromagnetic forces. Creation of a satisfactory quantum theory of gravity has been a very difficult problem and is unsolved to this day, though many people are working on it.

Chapter 4

Molecular Orbital Theory

PURPOSE

First, we define the problem, beginning with the Schrödinger equation

$$\hat{H}\psi = E\psi$$

Our goal is to come up with an analytic equation for the energy which can be minimized with respect to some variational parameter(s) to give an upper bound on the energy. To do this, we must

1. Understand the hamiltonian operater $\hat{H}$.
2. Find an appropriate wave function ψ which allows simple calculation of the electronic energy.
3. Examine potential variational parameters to figure out how to minimize the energy.

It will likely be convenient to have an outline of where we're going, so here it is:

1. Puropse (which we've been through)
2. The Hamiltonian, $\hat{H}$
   - Kinetic energy operators
   - Coulombic potential operators
3. The Wave Function, ψ
   - 1 electron orbitals

- Hartree products
- Slater determinants

4. Hamiltonian as Energy Operator, Schrödinger Equation
   - $E = \langle \psi | \hat{H} | \psi \rangle$
   - Integrals over one electron operators
   - Integrals over two electron operators
   - Specific case energy expressions
   - General form of HF energy
5. What Variational Parameter?
   - LCAO-MO theory, energy in AO basis
   - Density Matrices
6. Hartree-Fock Equations
   - Lagrangs's Undetermined Multipliers
   - A load 'o math
7. Matrix Formalism
8. Programme Outline

THE HAMILTONIAN, $\hat{H}$

The Hamiltonian is the total energy operator for a system, and is written as the sum of the kinetic energy of all the components of the system and the internal potential energy. In an atom or molecule, comprised of positive nuclei and negative electrons, the potential energy is simply that due to the coulombic interactions present.

Thus for the kinetic energy in a system of M nuclei and N electrons:

$$\hat{T}_N = -\sum_{A}^{M} \frac{1}{2M_A} \nabla_A^2$$

$$\hat{T}_e = -\sum_{i}^{N} \frac{1}{2} \nabla_i^2$$

And for the potential energy:

$$\hat{V}_{NN} = \sum_{A>B}^{M} \frac{Z_A Z_B}{r_{AB}}$$

$$\hat{V}_{ee} = \sum_{i>j}^{N} \frac{1}{r_{ij}}$$

$$\hat{V}_{eN} = -\sum_{A}^{M}\sum_{i}^{N} \frac{Z_A}{r_{iA}}$$

$$\hat{H} = \hat{T} + \hat{V}$$

Since,

$$\hat{H} = -\sum_{i}^{N} \frac{1}{2}\nabla_i^2 - \sum_{A}^{M}\sum_{i}^{N} \frac{Z_A}{r_{iA}} + \sum_{i>j}^{N} \frac{1}{r_{ij}} - \sum_{A}^{N} \frac{1}{2M_A}\nabla_A^2 + + \sum_{A>B}^{M} \frac{Z_A Z_B}{r_{AB}}$$

Within the Born-Oppenheimer approximation, we assume the nuclei are held fixed while the electrons move really fast around them. (note: $M_p / M_e \approx 1840$.) In this case, nuclear motion and electronic motion are seperated. The last two terms can be removed from the total hamiltonian to give the electronic hamiltonian, $\hat{H}_e$, since $\hat{V}_{NN} = K$, and $\nabla_A^2 = 0$. The nuclear motion is handled in a rotational/vibrational analysis. We will be working within the B-O approximation, so realizing

$$\hat{H}_e = -\sum_{i}^{N} \frac{1}{2}\nabla_i^2 - \sum_{A}^{M}\sum_{i}^{N} \frac{Z_A}{r_{iA}} + \sum_{i>j}^{N} \frac{1}{r_{ij}}$$

we completly define the problem. Solving the electronic Schrödinger equation using this will give the electronic structure of a molecular system at a fixed nuclear geometry.

THE WAVE FUNCTION, ψ

We've derived a complete many-electron Hamiltonian operator. Of course, the Schrödinger equation involving it is intractable, so let's consider a simpler problem, involving the one-electron hamiltonian

$$\hat{h}(i) = -\frac{1}{2}\nabla_i^2 + \sum_A^M \frac{1}{r_{iA}}$$

which involves no electron-electron interaction. This soluable in the B-O approximation (recall the hydrogen atom by letting M=1). Call the solutions to the one-electron Schrödinger equation $X_i(x)$. These will be molecular spin orbitals when we get around to it, but for now let it suffice to know they satisify the eigenequation

$$\hat{h}(i)X_j(x_i) = \varepsilon_j X_j(x_i)$$

with the interpretation that electron i occupies spin orbital x_j with energy ε_j. If we ignore electron-electron interaction in $\hat{H}_e$, we construct a simpler system with Hamiltonian

$$\hat{H} = \sum_i^N \hat{h}(i)$$

It will have eigenfunctions which are simple products of occupied spin orbitals, and thus an energy which is a sum of individual orbital energies, as

$$\psi^{HP} = X_i(x_1)X_j(x_2)X_k(x_3)...X_\eta(x_N)$$

$$\langle\psi|\hat{H}|\psi\rangle = \varepsilon_i + \varepsilon_j + \varepsilon_k + ... + \varepsilon_\eta = E$$

This kind of wavefunction is called a Hartree Product, and it is not physically realistic. In the first place, it is an independent-electron model, and we know electrons repel each other. Secondly, it does not satisfy the antisymmetry principle due to Pauli which states that the sign of the wavefunction must be inverted under the operation of switching the coordinates of any two electrons, or

$$\psi(...X_i...X_j...) = -\psi(...X_j...X_i...)$$

Part of the proof of equation acknowledges this is not so for a Hartree Product. To remedy this, first consider a two-electron system, such as helium. Two equivalent Hartree Product wavefunctions for this system are

$$\psi_1^{HP} = X_i(x_1)X_j(x_2)$$
$$\psi_2^{HP} = X_i(x_2)X_j(x_1)$$

Obviously, neither of these is appropriate. However, using the old "by inspection..." trick, we notice that

$$\psi = \frac{1}{\sqrt{2}}\left[X_i(x_1)X_j(x_2) - X_i(x_2)X_j(x_1)\right]$$

does. The mathematical form of this wavefunction can be generated by a determinant of 's,

$$\psi = 2^{-1/2}\begin{vmatrix} X_i(x_1) & X_j(x_1) \\ X_i(x_2) & X_j(x_2) \end{vmatrix}$$

The familiar Pauli exclusion principle follows directly from this example. When we attempt to doubly occupy a spin orbital X_i by putting electron 1 and electron 2 in it, what happens?

$$\psi = 2^{-1/2}\begin{vmatrix} X_i(x_1) & X_i(x_1) \\ X_i(x_2) & X_i(x_2) \end{vmatrix}$$

$$= \frac{1}{\sqrt{2}}\left[X_i(x_1)X_i(x_2) - X_i(x_2)X_i(x_1)\right]$$

$$= 0$$

Equation can be generalized to give the N electron Slater determinant

$$\psi = (N!)^{-1/2}\begin{vmatrix} X_i(x_1) & X_j(x_1) & \cdots & X_\eta(x_1) \\ X_i(x_2) & X_j(x_2) & \cdots & X_\eta(x_2) \\ \vdots & \vdots & \ddots & \vdots \\ X_i(x_N) & Xj(x_N) & \cdots & X_\eta(x_N) \end{vmatrix}$$

A shorthand notation for a Slater determinant has been introduced, where all the diagonal elements in the determinant are written in order as a "ket" vector. Equation can thus be written as

$$\psi = \left| X_i(x_1) X_j(x_2) \ldots X_\eta(x_n) \right|$$

where the normalization constant is absorbed into the notation.

Now we have written down a wave function appropriate for use in the case where $\hat{H} = \sum \hat{h}(i)$. In HF theory, we make some simplifications so many-electron atoms and molecules be treated this way. By tacitly assuming that each electron moves in a percieved electric field generated by the stationary nuclei and the average spatial distribution of all the other electrons, it essentially becomes an independant-electron problem. The HF Self Consistent Field procedure (SCF) will be bent on constructing each $X(x)$ to give the lowest energy.

ENERGY EXPRESSIONS

Let's assume a wave function of the Slater determinant form and find an expression for the expectation value of the energy. We've written a Slater determinant as a ket vector in shorthand notation, allowing us to make use of Dirac notation for such things as overlap. In this context, recall that

$$\begin{aligned} \langle \psi_a | \psi_b \rangle &= (\psi_a, \psi_b) \\ &= \int \psi_a^*(x) \psi_b(x) dx \end{aligned}$$

where the basis vectors ψ is expanded in are every possible value of x with contraction coefficients identified as the value of $\psi(x)$ at x. Thus placing an operator (such as $\hat{H}$) inside the bracket, we get the expectation value of the observable associated with that operator. Since $\hat{H}$ is the energy operator,

$$\begin{aligned} E_0 &= \int d\tau \psi_0^* \hat{H}_e \psi_0 \\ &= \left\langle \psi_0 \middle| \hat{H}_e \middle| \psi_0 \right\rangle \end{aligned}$$

is the differential of all the spin and space coordinates of all the electrons.

With much foresight, we continue to simplify the problem by writing $\hat{H}$ as a sum of one- and two-electron operators

$$\hat{H}_e = \sum_i \hat{h}(i) + \sum_{i>j} \frac{1}{r_{ij}}$$

$$= \hat{H}_1^{core} + \hat{H}_2$$

This will allow us to more precisely develop the electronic energy by it's components.

First, examine the core hamiltonian $\hat{H}_1^{core}$.

$$\langle \psi | \hat{H}_1^{core} | \psi \rangle = \sum_i \langle \psi | \hat{h}(i) | \psi \rangle$$

$$= \sum_i \langle X_1(x_1) \ldots . X_i(x_i) \ldots | \hat{h}(i) | X_1(x_1) \ldots X_i(x_i) \ldots \rangle$$

The nature of this is best evidenced by example, so we turn to the familiar helium atom, $\psi = |X_1(x_1) X_2(x_2)\rangle$. Look at one term in the above sum, for the sake of illustration take . $\langle X_1(x_1) X_2(x_2) | \hat{h}(1) | X_1(x_1) X_2(x_2) \rangle$

$$= \frac{1}{2} \int \left[X_1^*(x_1) X_2^*(x_2) - X_1^*(x_2) X_2^*(x_1) \right] \hat{h}(1)$$

$$\left[X_1(x_1) X_2(x_2) - X_1(x_2) X_2(x_1) \right] dx_1 dx_2$$

$$= \frac{1}{2} \int X_1^*(x_1) X_2^*(x_2) \hat{h}(1) X_1(x_1) X_2(x_2) d\tau$$

$$+ \frac{1}{2} \int X_1^*(x_2) X_2^*(x_1) \hat{h}(1) X_1(x_2) X_2(x_1) d\tau$$

$$- \int X_1^*(x_2) X_2^*(x_1) \hat{h}(1) X_1(x_1) X_2(x_2) d\tau$$

$$- \int X_1^*(x_1) X_2^*(x_2) \hat{h}(1) X_1(x_2) X_2(x_1) d\tau$$

$$= \frac{1}{2} \left(\langle X_1 | \hat{h} | X_1 \rangle + \langle X_2 | \hat{h} | X_2 \rangle \right)$$

Here $\langle X_1 | \hat{h} | X_1 \rangle$ is defined as $\int X_1^*(x_1)\hat{h}(1)X_1(x_1)dx_1$. In the first two terms of equation, the integrations over electron two's coordinates can be carried out irrespective of electron one's, and give the two terms of equation. The last two terms integrate to zero due to the orthogonality of X_1 and X_2. Repeating this for $\hat{h}(2)$ we get exactly the same thing, and we see

$$\sum_i \langle \psi | \hat{h}(i) | \psi \rangle = \langle X_1 | \hat{h} | X_1 \rangle + \langle X_2 | \hat{h} | X_2 \rangle$$

Profound, isn't it? Seems that every occupied spin orbital X_i yields a term of the form $\langle X_i | \hat{h} | X_i \rangle$ to the one electron energy.

Now look at $\hat{H}_2$.

$$\langle \psi | \hat{H}_2 | \psi \rangle = \sum_{i>j} \left\langle \psi \left| \frac{1}{r_{ij}} \right| \psi \right\rangle$$

$$= \sum_{i>j} \left\langle X_1(x_1) \ldots X_i(x_i) \ldots \left| \frac{1}{r_{ij}} \right| X_1(x_1) \ldots X_i(x_i) \ldots \right\rangle$$

Continuing to work in the helium atom example (realise that this could be two electron system) pick $(i,j) = (1,2)$ and look at that one term.

$$\left\langle X_1(x_1) X_2(x_2) \left| \frac{1}{r_{12}} \right| X_1(x_1) X_2(x_2) \right\rangle$$

$$= \frac{1}{2} \iint \left[X_1(x_1) X_2(x_2) - X_1(x_2) X_2(x_1) \right] \frac{1}{r_{12}}$$

$$\left[X_1(x_1) X_2(x_2) - X_1(x_2) X_2(x_1) \right] dx_1 dx_2$$

$$= \frac{1}{2} \int X_1^*(x_1) X_2^*(x_2) \frac{1}{r_{12}} X_1(x_1) X_2(x_2) d\tau$$

$$+\int X_1^*(x_2)X_2^*(x_1)\frac{1}{r_{12}}X_1(x_2)X_2(x_1)d\tau$$

$$-\int X_1^*(x_2)X_2^*(x_1)\frac{1}{r_{12}}X_1(x_1)X_2(x_2)d\tau$$

$$-\int X_1^*(x_1)X_2^*(x_2)\frac{1}{r_{12}}X_1(x_2)X_2(x_1)d\tau$$

Unfortunately, the $\frac{1}{r_{12}}$ operator prevents seperation of the integrations over the electronic coordinates of electron 1 and electron 2. It cannot be assured that the last two terms are zero. In general, they are not. However, since the x_1 and x_2 are dummy variables, the first and second terms of equation are equal, as are the last two. Thus for the two electron operator $\frac{1}{r_{12}}$,

$$\left\langle X_1(x_1)X_2(x_2)\left|\frac{1}{r_{12}}\right|X_1(x_1)X_2(x_2)\right\rangle$$

$$= \langle 12|12\rangle - \langle 12|21\rangle = \langle 12\|12\rangle$$

where

$$\langle ij\,|\,kl\rangle = \int dx_1 dx_2 X_i^*(x_1)X_j^*(x_2)\frac{1}{r_{12}}X_k(x_1)X_1(x_2)$$

$$\langle ij\,|\,kl\rangle = \langle ij\,|\,kl\rangle - \langle ij\,|\,lk\rangle$$

The construction $\langle ij\,\|\,kl\rangle$ is called an antisymmetrized two electron integral in physicists notation.

By working in the spin orbital basis, much trouble is avoided. In fact, by extending the results shown previously to the general case, we can now write down the HF energy for a given set of occupied spin orbitals.

$$E_{HF} = \left\langle \psi\left|\hat{H}_e\right|\psi\right\rangle = \left\langle \psi\left|\hat{H}_1^{core} + \hat{H}_2\right|\psi\right\rangle$$

$$E_{HF} = \langle X_1(x_1)...X_i(x_i)...|\hat{H}_1|X_1(x_1)...X_i(x_i)...\rangle +$$
$$\langle X_1(x_1)...X_i(x_i)...|\hat{H}_2|X_1(x_1)...X_i(x_i)...\rangle$$
$$= \sum_i \langle X_1(x_1)...X_i(x_i)...|\hat{h}(i)|X_1(x_1)...X_i(x_i)...\rangle +$$
$$\sum_{i>j} \left\langle X_1(x_1)...X_i(x_i)...\left|\frac{1}{r_{12}}\right|X_1(x_1)...X_i(x_i)...\right\rangle$$
$$= \sum_i \langle X_i|h|X_i\rangle + \sum_{i>j} \langle ij \| ij\rangle$$

Now move on and consider working in the spatial orbital basis, where

$$X_i(x) = \psi_a(r)w$$

This is more natural, since our intuition is usually based on having a region of space which describes the location (more or less) of electrons, one of alpha spin and one of beta spin. Some of quantum chemistry is formulated entirely in terms of spin orbitals, for various reasons. For our purposes, we will work entirely in the spatial orbital basis. This will cause things to get somewhat murky soon, but in the long run it will be simpler.

At any rate, in the two electron system we adore so much, we can identify the two occupied spin orbitals X_1 and X_2 as the spin up and spin down halves of the single lowest lying spatial orbital, a 1s in helium or the bonding orbital in H^2 for example. These can be more precisely defined as

$$X_1(x) = \psi_1(r)\alpha$$
$$X_2(x) = \psi_1(r)\beta$$

This changes the way we write slater determinants. Using an overbar to denote β spin occupation of a spatial orbital, $\overline{\psi}$,

$$|X_1(x_1)X_2(x_2)\rangle = |\psi_1(r_1)\overline{\psi}_2(r_2)\rangle = |\psi_1\overline{\psi}_1\rangle$$

Rethinking the one electron integrals for this case,

$$\sum_i \langle \psi_1\bar{\psi}_1 \,|\, \hat{h}(i) \,|\, \psi_1\bar{\psi}_1 \rangle = \langle \psi_1 | \hat{h}(1) | \psi_1 \rangle + \langle \bar{\psi}_1 | \hat{h}(2) | \bar{\psi}_1 \rangle$$

$$= (\psi_1 | \hat{h} | \psi_1) + (\psi_1 | \hat{h} | \psi_1)$$

$$\equiv h_{11} + h_{11}$$

The notation $(\psi_1 | \hat{h} | \psi_1)$ is used to denote an integral over only spatial coordinates, what remains after the spin integrations have been carried out, giving a factor of 1 or 0.

That was a neat closed shell system. How about something like $\psi = |\psi_1\bar{\psi}_1\psi_2\rangle$?

$$\sum_i \langle \psi | \hat{h}(i) | \psi \rangle = \langle \psi_1 | \hat{h} | \psi_1 \rangle + \langle \bar{\psi}_1 | \hat{h} | \bar{\psi}_1 \rangle + \langle \psi_2 | \hat{h} | \psi_2 \rangle$$

$$= (\psi_1 | \hat{h} | \psi_1) + (\psi_1 | \hat{h} | \psi_1) + (\psi_2 | \hat{h} | \psi_2)$$

$$= 2h_{11} + h_{22}$$

$$= 2\sum_i^{docc} h_{ii} + \sum_i^{socc} h_{ii}$$

$$= 2\sum_i^{occ} f_i h_{ii}$$

The coefficient f_i here is related to the occupation of spatial orbital i, and will be more precisely defined later. The two electron integrals are a tad more involved, but we go about it in essentially the same manner.

$$\langle X_1X_2 | X_1X_2 \rangle = \langle \psi_1\bar{\psi}_1 | \psi_1\bar{\psi}_1 \rangle$$

$$= [\psi_1\psi_1 | \bar{\psi}_1\bar{\psi}_1]$$

$$= (\psi_1\psi_1 | \psi_1\psi_1)$$

$$\langle X_1X_2 | X_2X_1 \rangle = \langle \psi_1\bar{\psi}_1 | \bar{\psi}_1\psi_1 \rangle$$

$$= [\psi_1\bar{\psi}_1 | \bar{\psi}_1\psi_1]$$
$$= \hat{U}$$

Yes, I know. Very confusing. But it's all just notation, and can be understood. In physicist's notation (equivalent to Dirac notation), $\langle \psi_i\psi_j | \psi_k\psi_l \rangle$ refers to the two electron integral where ψ_i and ψ_k are functions of electron 1, while ψ_j and ψ_l are functions of electron 2. Chemist's notation (with the square brackets []) places the functions of electron 1 on the left and the functions of electron 2 on the right. When the two functions of a single electron are not of the same spin, the whole integral goes to zero, otherwise the spin integrates out to 1. Hence the spin-free notation $(\psi_1\psi_1 | \psi_1\psi_1)$. What occurs when the two electrons are of parallel spin, requiring distinct spatial orbitals and a wavefunction something like $\psi = |\psi_1\psi_2\rangle$? The same general form is present, and a related antisymmetrized two electron integral is evaluated. In this case, $\langle \psi_1\psi_2 \| \psi_1\psi_2 \rangle$.

$$\langle \psi_1\psi_2 \| \psi_1\psi_2 \rangle = \langle 12 \| 12 \rangle$$
$$= \langle 12 | 12 \rangle - \langle 12 | 12 \rangle$$
$$= [11 | 22] - [12 | 21]$$
$$= (11 | 22) - (12 | 21)$$
$$\langle \psi_1\psi_2 \| \psi_1\psi_2 \rangle = J_{12} - K_{12}$$

Another bit of notation, which should be apparent. $J_{12} = (11 | 22)$ and $K_{12} = (12 | 21)$. J_{ij} is termed a coulomb integral and has the physically reassuring interpretation of somehow accounting for electronic repulsion between electrons in molecular orbital *i* and molecular orbital *j*. K_{ij}, the exchange integral, has no classical analog and no true physical interpretation. Many have tried to come up with something, and a typical attempt says that it "correlates the motions of electron i and j when they have parallel spins, lowering the energy since those electrons avoid each other better." Whatever.

Best to just move on to a general energy equation in the spatial MO basis. In summary:

- The one electron integrals contribute h_{ii} for each electron in orbital i.
- The two electron integrals contribute J_{ij} for each pair of electrons, and $\left(-K_{ij}\right)$ for each pair of parallel spin.

$$E_{HF} = 2\sum_{i}^{occ} f_i h_{ii} + \sum_{i,j}^{occ} \left\{ \alpha_{ij}\left(ii \mid jj\right) + \beta_{ij}\left(ij \mid ij\right) \right\}$$

$$= 2\sum_{i}^{occ} f_i h_{ii} + \sum_{i,j}^{occ} \left\{ \alpha_{ij} J_{ij} + \beta_{ij} K_{ij} \right\}$$

$$f_i = \begin{cases} 1 \text{ if } i \text{ doubly occupied} \\ 1/2 \text{ if } i \text{ singly occupied} \end{cases}$$

$$\alpha_{ij} = \begin{cases} 2 \text{ if } i \text{ and j doubly occupied} \\ 1 \text{ if } i \text{ or } j \text{ doubly occupied and the other singly occupied} \\ 1/2 \text{ if } i \text{ and } j \text{ singly occupied} \end{cases}$$

$$\alpha_{ij} = \begin{cases} -1 \text{ if } i \text{ and } j \text{ doubly occupied} \\ -1/2 \text{ if } i \text{ or } j \text{ singly occupied, the other doubly occupied} \\ -1/2 \text{ if } i \text{ and } j \text{ singly occupied, with parallel spin} \\ 1/2 \text{ if } i \text{ and } j \text{ singly occupied, with opposite spin} \end{cases}$$

WHAT VARIATIONAL PARAMETER?

Ah, the crux of the problem, is it not? Up until now, we've just assumed we have some set of molecular orbitals X_i or ψ_i, which we can manipulate at will. But how does one come up with even approximate solutions to the many body Schrödinger equation without having to solve it? Start with the celebrated linear combination of atomic orbitals to get molecular orbitals (LCAO-MO) approximation. This allows us to use some set of (approximate) atomic orbitals, the basis functions which we know and love, to expand the MOs in. In the most general terms,

$$\psi_i = \sum_{\mu} C^i_{\mu}\phi_{\mu}$$

ψ_i remains a spatial molecular orbital, ϕ_μ is a spatial atomic orbital (perhaps symmetry orbital, but no matter), and C^i_μ are the contraction coefficients by which we transform from one basis to another. Armed with only this, we should be able to compose the electronic energy in the atomic orbital basis. Why, you ask? Because we have an expression in terms of integrals over MOs.

To variationally minimize that energy, we need to vary the MOs themselves, but have no way to do that, since they remain these amorphous constructs. By defining them a bit more precisely, we should arrive at a point where an obvious set of variational parameters (hint: C^i_μ) present themselves. Begin with the closed shell HF energy in terms of spatial MOs.

$$\begin{aligned} E_{HF} &= 2\sum_{i}^{docc} h_{ii} + \sum_{i,j}^{docc} \{2J_{ij} - K_{ij}\} \\ &= 2\sum_{i}^{docc} h_{ii} + \sum_{i,j}^{docc} \{2(ii\,|\,jj) - (ij\,|\,ij)\} \end{aligned}$$

Since $(\psi_i|h|\psi_i) = \int \psi_i^*(r_1)h(1)\psi_i(r_1)\,dr_1$ and

$\psi_i = \sum_\mu C^i_\mu \phi_\mu$,

$$\int \phi_i^*(r_1)h(1)\psi_i(r_1)\,dr_1 =$$

$$\int \sum_{\mu} C^{i*}_{\mu}\phi^*_{\mu}(r_1)h(1)\sum_{\nu} C^i_{\nu}\phi_{\nu}(r_1)\,dr_1$$

$$= \sum_{\mu}\sum_{\nu} C_{\mu}^{i*} C_{\nu}^{i} \int \phi_{\mu}^{*}(r_1) h(1) \phi_{\nu}(r_1) dr_1$$

$$= \sum_{\mu}\sum_{\nu} C_{\mu}^{i*} C_{\nu}^{i} (\phi_{\mu} | h | \phi_{\nu})$$

$$= \sum_{\mu}\sum_{\nu} C_{\mu}^{i*} C_{\nu}^{i} h_{\mu\nu}$$

$(\phi_{\mu} | h | \phi_{\nu})$ is a one electron integral over atomic orbitals. Do we have something we can actually calculate!? Take an aside and examine this quantity superficially.

Typically, basis functions are constructed to mimic true atomic orbitals. The Hydrogen atom can be described rigorously, and the eigenfunctions found. The 1s orbital looks something like $Ne^{-\zeta r}$. It satisfies all the appropriate boundary conditions, having a cusp at the nucleus and exponentially decaying to zero at infinity. Higher angular momentum functions, like 2p's and 3d's, can be built from this basic framework through adding the angular nodes by multiplying in factors of x, y, and z. Basis functions such as these are called slater-type orbitals. If instead of the exponential $Ne^{-\zeta r}$ we use a gaussian function, $Ne^{-\alpha r^2}$, we loose the boundary conditions but generate a more tractable problem when it comes to calculating integrals. Using a linear combination of single cartesian gaussian-type orbitals to approximate a slater-type orbital gives better computational accuracy without too much more effort. Here's the functional forms of all three types:

$$\phi_{\mu}^{STO}(r) = Nx^{1} y^{m} z^{n} e^{-\zeta r}$$

$$\phi_{\nu}^{SGTO}(r) = Nx^{1} y^{m} z^{n} e^{-\alpha r^2}$$

$$\phi_{\mu}^{CGTO}(r) = \sum_{\nu} C_{\nu}^{\mu} \phi_{\nu}^{SGTO}(r)$$

Just taking a 1s SGTO for illustrative purposes, what is that one electron integral?

$$\left(\phi_{1s}^{SGTO}(r_1)\middle|h(1)\middle|\phi_{1s}^{SGTO}(r_1)\right)= N^2\int e^{-ar^2}\left(-\nabla^2+\frac{1}{r}\right)e^{-ar^2}\,dr$$

Hey! We can do that!

Back to the problem at hand, we now need to expand the two electron integrals in the MO basis. Following a procedure analogus to equation, we get

$$(ij\mid kl) = \sum_\mu\sum_\nu\sum_\rho\sum_\sigma C_\mu^{i*}C_\nu^{j*}C_\rho^{k}C_\sigma^{l}(\mu\nu\mid\rho\sigma)$$

All of these AO integrals can be calculated and stored, to be called up when needed to evaluate the electronic energy. The closed shell energy in the AO basis can be written as

$$E^{\text{clscf}} = 2\sum_{\mu\nu}D_{\mu\nu}h_{\mu\nu} + \sum_{\mu\nu\rho\sigma}D_{\mu\nu}D_{\rho\sigma}\left\{2(\mu\nu\mid\rho\sigma)-(\mu\rho\mid\nu\sigma)\right\}.$$

$$= \sum_{\mu\nu}D_{\mu\nu}\left[2h_{\mu\nu}+\sum_{\rho\sigma}D\rho\sigma\left\{2(\mu\nu\mid\rho\sigma)-(\mu\rho\mid\nu\sigma)\right\}\right]$$

$D_{\mu\nu}$ is the density matrix, a product of AO-MO coefficient matrices, or

$$D_{\mu\nu}=\sum_{\mu\nu}C_\mu^{i*}C_\nu^{i}$$

$$D_{\mu\nu}=C^{\dagger}C$$

HARTREE FOCK EQUATIONS

The electronic energy is a functional of the spin orbitals, and we want to minimize it subject to some set of constraints. This can be done using the calculus of variations applied to functionals. So lets look at a general example of functional variation applied to *E*, a functional of some trial wavefunction Φ that can be linearly varied under a single constraint.

$$E = \langle \Phi | H | \Phi \rangle$$

$$|\Phi\rangle = \sum_i^N C_i |\psi_i\rangle$$

By equation, we see that $E[\Phi]$, depending on the form of the wavefunction, and by equation that $|\Phi\rangle$ can be linearly expanded (hence linearly varied) in some set of N functions. This is directly analogous to expanding the asymmetric top rotational wavefunctions in a complete set of symmetric top rotational wavefunctions. The task is to minimize E subject to the single constraint that the wavefunction Φ remain normalized, or

$$\langle \Phi | \Phi \rangle = \sum_{ij} c_i^* c_j \langle \psi_i | \psi_j \rangle = 1$$

Writing the energy as

$$E = \langle \Phi | H | \Phi \rangle = \sum_{ij} c_i^* c_j \langle \psi_i | H | \psi_j \rangle$$

we want $\delta E = \hat{U}$, so

$$\delta E = \frac{\partial E}{\delta c_1}\delta c_1 + \frac{\partial E}{\delta c_2}\delta c_2 + \frac{\partial E}{\delta c_3}\delta c_3 + \ldots + \frac{\partial E}{\delta c_N}\delta c_N$$

$$= \sum_i^N \frac{\partial E}{\partial c_i}\delta c_i = 0$$

However, due to the normalization constraint, there is one linear dependancy in the expansion coefficients. If we simply solve equation for the coefficients, they may no longer be normalized, and if we solve for N-1 of the coefficients and invoke the normalization condition to determine the Nth, the energy may not be stationary about it. So we construct the function

$$L[\Phi] = \langle \Phi | H | \Phi \rangle + E\left(\langle \Phi | \Phi \rangle - 1\right)$$

$$L[\Phi] = \sum_{ij} c_i^* c_j \langle \psi_i | H | \psi_j \rangle + \sum_{ij} c_i^* c_j E\left(\langle \psi_i | \psi_j \rangle - 1\right)$$

Taking the differential of *L*,

$$\delta L[\Phi] = \sum_{ij} \delta c_i^* c_j \langle \psi_i | H | \psi_j \rangle + \sum_{ij} c_i^* \delta c_j \langle \psi_i | H | \psi_j \rangle +$$

$$E \sum_{ij} \delta c_i^* c_j \langle \psi_i | \psi_j \rangle + E \sum_{ij} c_i^* \delta c_j \langle \psi_i | \psi_j \rangle$$

$$= \sum_{ij} \delta c_i^* c_j \left[\langle \psi_i | H | \psi_j \rangle + E \langle \psi_i | \psi_j \rangle \right]$$

$$+ \sum_{ij} c_i^* \delta c_j \left[\langle \psi_i | H | \psi_j \rangle + E \langle \psi_i | \psi_j \rangle \right]$$

$$= \sum_{ij} \delta c_i^* \left[\sum_j c_j \left(\langle \psi_i | H | \psi_j \rangle + E \langle \psi_i | \psi_j \rangle \right) \right] +$$

$$\sum_{ij} \delta c_j \left[\sum_j c_i^* \left(\langle \psi_i | H | \psi_j \rangle + E \langle \psi_i | \psi_j \rangle \right) \right]$$

$$= \sum_i \delta c_i^* \left[\sum_j c_j \left(\langle \psi_i | H | \psi_j \rangle + E \langle \psi_i | \psi_j \rangle \right) \right] +$$

$$\sum_i \delta c_i \left[\sum_j c_j^* \left(\langle \psi_j | H | \psi_i \rangle + E \langle \psi_j | \psi_i \rangle \right) \right]$$

$$= \sum_i \delta c_i^* \left[\sum_j c_j \left(H_{ij} + ES_{ij} \right) \right] + \sum_i \delta c_i \left[\sum_j c_j^* \left(H_{ij}^* + ES_{ij}^* \right) \right] = 0$$

Since δc_i and δc_i^* are arbitrary, the bracketed parts of equation must be zero. Thus

$$\sum_j \left(H_{ij} c_j - ES_{ij} c_j \right) = 0$$

It is clear that this can be written as a matrix product, and is in fact an eigenvalue equation in the form

$$Hc = ScE$$

Knowing that H and S are hermetian, this matrix eigenvalue equation can be rewritten as

$$S^{-1/2}Hc = S^{1/2}cE$$

$$S^{-1/2}HS^{-1/2}S^{1/2}c = S^{1/2}cE$$

The matrix $S^{-1/2}\ HS^{-1/2}$ is symmetric and easily diagonalized, with eigenvectors $S^{1/2}c$. These can be transformed (multiply on the left by $S^{-1/2}$) to give the optimal coefficients c_i for stationary state. This is very powerful, since in one fell swoop we've got the entire energy spectrum and the appropriate wave functions, properly orthonormal, for all the states. This should illustrate the general technique we will be employing to develop the Hartree-Fock equations and from them the algebraic Roothaan equations, which you will be programming later this summer.

On to the true problem. Assume we have a wave function in the form of a Slater determinant of spin orbitals, $\psi_0 = |X_1X_2...X_N\rangle$. We state the problem as:

Please minimize the electronic energy of this single determinant subject to the constraint that the spin orbitals all remain orthonormal to one another.

We already understand the energy fine by equation, and the constraint can be simply stated as

$$\langle a\,|\,b\rangle - \delta_{ab} = 0$$

There are N spin orbitals, so there are $N(N+1)/2$ independent constraints (note: $\langle a\,|\,b\rangle = \langle b\,|\,a\rangle^*$), so we need that many undetermined multipliers in our lagrangian function, for which we present

$$L\left[\{X_a\}\right] = E_0\left[\{X_a\}\right] - \sum_a^N \sum_b^a \varepsilon_{ba}\left(\langle a\,|\,b\rangle - \delta_{ab}\right)$$

The restricted sum will prove inconvenient, but it can be eliminated. By taking the complex conjugate of the constraints and the lagrangian function, equation,

$$\langle a\,|\,b\rangle^* - \delta_{ab} = \hat{U}$$

$$\langle a|b\rangle - \delta_{ab} = \hat{U}$$

$$L[\{X_a\}] = E_0[\{X_a\}] - \sum_a^N \sum_b^a \varepsilon_{ba}^* (\langle b|a\rangle - \delta_{ab})$$

we realise we can unrestrict the summation by restraining ε_{ab} to be a hermitian matrix, such that $\varepsilon_{ab}^* = \varepsilon_{ba}$. This introduces no new undetermined multipliers into the equation, and creates a form more amenable to further derivation. Thus the Lagrangian function we will be working with is

$$L[\{X_a\}] = E_0[\{X_a\}] - \sum_a^N \sum_b^N \varepsilon_{ba} (\langle a|b\rangle - \delta_{ab})$$

The differential of this function must be set to zero as before, giving us

$$\delta L = \delta E_0 - \sum_a^N \sum_b^N \varepsilon_{ba} \delta \langle a|b\rangle = \hat{U}$$

Since we have an electronic energy in terms of spin orbitals

$$E_0 = \sum_a^N \langle a|H|a\rangle + \frac{1}{2} \sum_{ab}^N \langle ab \| ab\rangle$$

$$E_0 = \sum_a^N \langle a|H|a\rangle + \frac{1}{2} \sum_{ab}^N [\langle ab|ab\rangle - \langle ab|ba\rangle],$$

we can write the variance of the energy δE_0 as

$$\delta E_0 = \sum_a^N (\langle \delta_a|H|a\rangle + \langle a|H|\delta_a\rangle)$$

$$+ \frac{1}{2} \sum_{ab}^N \begin{bmatrix} \langle \delta ab|ab\rangle + \langle a\delta b|ab\rangle + \langle ab|\delta ab\rangle + \langle ab|a\delta b\rangle + \\ -\langle \delta ab|ba\rangle - \langle a\delta b|ba\rangle - \langle ab|\delta ba\rangle - \langle ab|b\delta a\rangle \end{bmatrix}$$

It takes some thought to realise that there are only two unique two electron integrals in this list, and tnat it can be written

$$\delta E_0 = \sum_{a}^{N} \langle \delta a | H | a \rangle + \sum_{ab}^{N} \left(\langle \delta ab \,|\, ab \rangle - \langle \delta ab \,|\, ba \rangle \right)$$

$$+ \text{complex conjugate}$$

The other variance we need is $\varepsilon_{ba}\delta\langle a \,|\, b\rangle$, which can be expanded as

$$\sum_{ab} \varepsilon_{ba} \langle \delta a \,|\, b \rangle + \sum_{ab} \varepsilon_{ba} \langle a \,|\, \delta b \rangle$$

$$= \sum_{ab} \varepsilon_{ba} \langle \delta a \,|\, b \rangle + \sum_{ab} \varepsilon_{ab} \langle b \,|\, \delta a \rangle$$

$$= \sum_{ab} \left(\varepsilon_{ba} \langle \delta a \,|\, b \rangle + \varepsilon_{ba}^{*} \langle \delta a \,|\, b \rangle^{*} \right)$$

So the whole thing boils down to one neat statement,

$$\delta L = \sum_{a} \langle \delta a | H | a \rangle + \sum_{a}^{N} \langle \delta a \,|\, b \rangle +$$

$$\sum_{ab}^{N} \left(\langle \delta ab \,\|\, ab \rangle - \varepsilon_{ba} \langle \delta a \,|\, b \rangle \right) + \text{complex conjugate}$$

If we conveniently define a coulomb operator $\hat{J}_b(1)$ and an exchange operator $\hat{K}_b(1)$ as

$$\hat{J}_b(1) = \int dx_2 \left| X_b(2) \right|^2 r_{12}^{-1}$$

$$\hat{K}_b(1) X_a(1) = \int dx_2 X_b^{*}(2) r_{12}^{-1} X_a(2) X_b(1)$$

we can rewrite the two electron integral

$$\langle ab \,\|\, ab \rangle = \int dx_1 X_a^{*}(1) \left(\hat{J}_b(1) - \hat{K}_b(1) \right) X_a(1)$$

$$= \left\langle a \middle| \left(\hat{J}_b(1) - \hat{K}_b(1) \right) \middle| a \right\rangle$$

This allows for a more compact notation to be employed in writing the variance in the lagrangian function

$$\delta L = \sum_{a} \int dx_1 \delta X_a^{*}(1)$$

$$\left[\hat{h}(1)X_a(1)+\sum_b\left\{\left(\hat{J}_a(1)-\hat{K}_b(1)\right)X_a(1)-\varepsilon_{ba}X_b(1)\right\}\right]$$

$$+complex\,conjugate=0$$

Like before, the part in brackets is forced to be zero, since $\delta X_a^*(1)$ can be anything. Setting it equal to zero and rearranging to make it look like some sort of eigenvalue equation yields

$$\left[\hat{h}(1)+\sum_b\left(\hat{J}_b(1)-\hat{K}_b(1)\right)\right]X_a(1)=\sum_b\varepsilon_{ba}X_b(1)$$

$$f(1)X_a(1)=\sum_b\varepsilon_{ba}X_b(1)$$

These are the glorious Hartree-Fock equations derived in general in the spin orbital basis. But wait - there's a problem. These are coupled integro-differential equations, and while they are not strictly unsolvable, they're a pain. It would be nice to at least uncouple them, so let's do that.

If we apply a unitary rotation to the full set of spin orbitals, generating a new set

$$X_a^1=\sum_b X_bU_{ba}$$

where U is unitary, ie. U† = U_{-1}, what changes? The rotation can be written as a matrix product, if we define A as the matrix resembling the slater determinant for the system, ie. $(N!)^{-1/2}\det(A)=|\psi_0\rangle$. In that case,

$$\det(A')=\det(A)\det(U)$$

However, since this matrix U is unitary, $\left|\det(U)^2\right|=1\det(U)=e^{i\phi}$, and the new wavefunction differs from the old by a phase factor, affecting nothing observable. How does it affect $f(1)$ and ε_{ba}?

$$\sum_a\hat{J}_a^1(1)=\sum_a\int dx_2X_a^{1*}(2)r_{12}^{-1}X_a^1(2)$$

$$= \sum_a \int dx_2 \sum_a U_{ba}^* X_b^*(2) r_{12}^{-1} \sum_c U_{ca} X_c(1)$$

$$= \sum_{bc} \left(\sum_a U_{ba}^* U_{ca} \right) \int dx_2 X_b^*(2) r_{12}^{-1} X_c(2)$$

$$= \sum_{bc} \delta_{bc} \int dx_2 X_b^*(2) r_{12}^{-1} X_c(2) = \sum_b \hat{J}_b(1)$$

$$\sum_a \hat{K}_a^1(1) X_{a'}^1(1) = \sum_a \int dx_2 X_b^{1*}(2) r_{12}^{-1} X_{a'}^1(2) X_a^1(1)$$

$$= \sum_a \int dx_2 \sum_b U_{ba}^* X_b^*(2) r_{12}^{-1} Xa'(2)' \sum_c U_{ca} X_c(1)$$

$$= \sum_{bc} \left(\sum_a U_{ba}^* U_{ca} \right) \int dx_2 X_b^*(2) r_{12}^{-1} Xa'(2)^1 X_c(1)$$

$$= \sum_{bc} \delta_{bc} \int dx_2 X_b^*(2) r_{12}^{-1} Xa'(2)^1 X_c(1) = \sum_b \hat{K}_b(1) X_{a'}^1(1)$$

So $f(1)' = f(1)$! How about ε_{ba}? Start by realizing that ε_{ba} are matrix elements of the fock operator.

$$\langle X_c | f | X_a \rangle = \sum_b \varepsilon_{ba} \langle X_c | X_b \rangle$$

$$= \varepsilon_{ca}$$

$$\varepsilon_{ab}^1 = \int dx_1 X_a^{l*} f(1) X_b^1(1)$$

$$= \sum_{cd} U_{ca}^* U_{a'b} \varepsilon_{cd}$$

$$= U^\dagger \varepsilon U$$

This last result can be written as a martix product as well, and it is seen that this is now a unitary transformation to the matrix . We are free to choose U to be whatever we please, and if we choose it to make diagonal, we can rewrite the Hartree-Fock equations as

$$f | X_a \rangle = \varepsilon_a | X_a \rangle$$

When this is done, the resulting spin orbitals are termed the Hartree-Fock canonical orbitals. Read section 3.3 in Szabo and Ostlund for various fun things to do with the Hartree-Fock equations.

The problem still remains, though. These are integro-differential equations, which computers (and computer programmers) balk at. That is why Roothaan is a HERO! Through his results, we can transform these into a set of matrix formulated algebraic equations that computers and programmers dig. The general case is too troublesome for now, so let's limit ourselves to closed shell, RHF orbitals. To take advantage of the simplifications this can afford, we need to return to the spatial orbital basis. We've derived the spin orbital based Fock operator

$$f(x_1) = \hat{h}(x_1) + \sum_a^N \left(\hat{J}_a(x_1) - \hat{K}_a(x_1) \right)$$

Without further ado, we'll introduce the spatial orbital based Fock operator and be done with it.

$$f(r_1) = \hat{h}(r_1) + \sum_a^{N/2} \left(2\hat{J}_a(r_1) - \hat{K}_a(r_1) \right)$$

$$\hat{J}_b(r_1) = \int dr_2 \left|\psi_b(r_2)\right|^2 r_{12}^{-1}$$

$$\hat{K}_a(r_1)\psi_a(r_1) = \int dr_2 \psi_b^*(r_2) r_{12}^{-1} \psi_a(r_2)\psi_b(1)$$

With the properties

$$f(r_1)\psi_j(r_1) = \varepsilon_j \psi_j(r_1)$$

$$\varepsilon_i = h_{ii} + \sum_b^{N/2} 2J_{ib} - K_{ib}$$

Now we introduce a basis set expansion to bring the HF integro-differential equations to soluable algebraic equations. Letting $\psi_i = \Sigma_\mu C_\mu^i \phi_\mu$,

$$f(1)\sum_{v} C_v^i \phi_v = \varepsilon_i \sum_{v} C_v^i \phi_v$$

Multiplying by $\phi_\mu^*(1)$ and integrating over electron 1 gives

$$\sum_{v} C_v^i \int dr_1 \phi_\mu^*(1) f(1) \phi_v(1) = \varepsilon_i \sum_{v} C_v^i \int dr_1 \phi_\mu^*(1) \phi_v(1)$$

Identifying the integrals as matrix elements of the fock operator and the unit operator (overlap) respectively,

$$\sum_{v} F_{\mu v} C_v^i = \varepsilon_i \sum_{v} S_{\mu v} C_v^i$$

Using the fact that is diagonal, this can be written as the matrix product

$$FC = SC\varepsilon$$

So what is F, this so called Fock Matrix? We've defined $F_{\mu v}$ before as the matrix element of the one electron fock operator, *f*(1) in equation. Writing out this integral and expanding,

$$F_{\mu v} = \int dr_1 \phi_\mu^*(1) f(1) \phi_v(1)$$

$$= \int dr_1 \phi_\mu^*(1) \left[h(1) + \sum_{a} \left(2\hat{J}_a(1) - \hat{K}_a(1) \right) \right] \phi_v(1)$$

$$= \int dr_1 \phi_\mu^*(1) h(1) \phi_v(1) +$$

$$\sum_{a} \left[2 \int dr_1 \phi_\mu^*(1) \hat{J}_a(1) \phi_v(1) - \int dr_1 \phi_\mu^*(1) \hat{K}(1) \phi_v(1) \right]$$

$$= h_{\mu v} + \sum_{a} 2(\mu v | aa) - (\mu a | av)$$

$$= h_{\mu v} + \sum_{a} \left[\sum_{\rho\sigma} C_\rho^a C_\sigma^a 2(\mu v | \rho\sigma) - (\mu\rho | \sigma v) \right]$$

$$= h_{\mu v} + \sum_{\rho\sigma} \left(\sum_{a} C_\rho^a C_\sigma^a \right) \left[2(\mu v | \rho\sigma) - (\mu\rho | \sigma v) \right]$$

$$= h_{\mu v} + \sum_{\rho\sigma} D_{\rho\sigma}\left[2(\mu v \mid \rho\sigma) - (\mu\rho \mid \sigma v)\right]$$

This is a quantity which can be easily constructed given a set of molecular orbitals (the coefficients C_{μ}^{a}) and a precalculated set of atomic orbital integrals. At this point, the Hartree-Fock equations have been reduced to a matrix eigenvector problem, $FC = SC\varepsilon$, but not in a computationally convenient form. Following the analysis leading to equation, we first define the transformed Fock matrix as

$$F' = S^{-1/2} F S^{-1/2}$$

We can then take

$$F'\left(S^{1/2}C\right) = \left(S^{1/2}C\right)\varepsilon$$

to be equivalent to equation. Since is diagonal by choice, diagonalizing the transformed Fock matrix gives a set of transformed molecular orbital coefficients, $C^{\dagger} = S^{1/2}C$. The matrix $S^{1/2}C$ can be back transformed to give the true MO coefficient matrix, C. The density matrix for these coefficients is formed by the product $D = C^{\dagger}C$, and can subsequently be used to construct a new fock matrix via equation. Since the overlap matrix S does not depend on the MO coefficients, the same unitary transformation can be applied to the new fock matrix to give a new transformed fock matrix. This can be diagonalized to produce new MO coefficients, and the process repeated until convergance. As an initial guess for the fock matrix, one generally uses the core hamiltonian, ignoring all the two electron integrals.

$$F_{\mu v}^{1} = H_{\mu v} = \int \phi_{\mu}^{*}\left(-\nabla_i^2 + \sum_{A} \frac{Z_A}{r_{iA}}\right)\phi_v$$

From the core hamiltonian, an initial C is obtained and a much improved Fock matrix can be built including the two electron integrals. And that's about it for the Hartree-Fock Self Consistent Field Method. These last few pages will be the most important when you get around to programming the closed shell SCF method for the specific case of water, as you will be given the integrals in a file, and you can begin the process by building the core hamiltonian as described above.

MATTER WAVES

We begin our study of quantum mechanics by discussing the diffraction undergone by X-rays and electrons when they interact with a crystal. X-rays are a form of electromagnetic radiation with wavelengths comparable to the distances between atoms. Scattering from atoms in a regular crystalline structure results in an interference pattern which is in many ways similar to the pattern from a diffraction grating. We first develop Bragg's law for diffraction of X-rays from a crystal. Two practical techniques for doing X-ray diffraction are then described.

It turns out that electrons have wave-like properties and also undergo Bragg diffraction by crystals. Bragg diffraction thus provides a crucial bridge between the worlds of waves and particles. With this bridge we introduce the classical ideas of momentum and energy by relating them to the wave vector and frequency of a wave. The properties of waves also give rise to the Heisenberg uncertainty principle.

This gives us a feel for the chronology of these discoveries and indicates how important they were to the development of physics in the early 20th century.

BRAGG'S LAW

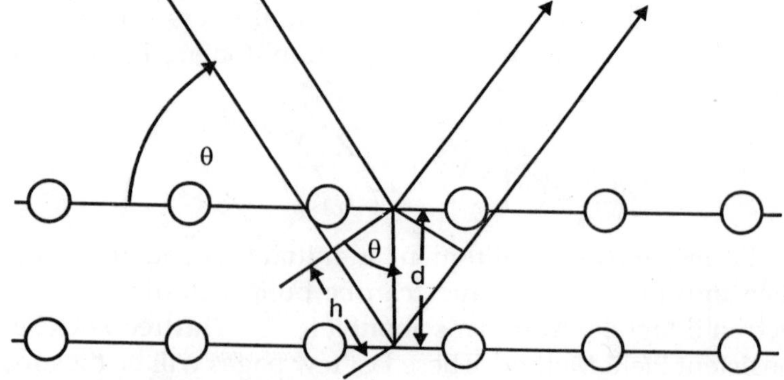

Fig. Schematic diagram for determining Bragg's law.

Figure schematically illustrates interference between waves scattering from two adjacent rows of atoms in a crystal. The net effect of scattering from a single row is equivalent to partial reflection from a mirror imagined to be aligned with the row. Thus, the angle of "reflection" equals the angle of incidence for each row. Interference then occurs between the beams reflecting off different rows of atoms in the crystal.

For the two adjacent rows shown in figure, the path difference between beams is $2h = 2d\sin\theta$. For constructive interference this must be an integer number of wavelengths, $m\lambda$, where the integer m is called the order of interference. The result is Bragg's law of diffraction:

$$m\lambda = 2d\sin\theta, m = 1,2,3.... \text{ (Bragg's law)}$$

If only two rows are involved, the transition from constructive to destructive interference as θ changes is gradual. However, if interference from many rows occurs, then the constructive interference peaks become very sharp with mostly destructive interference in between. This sharpening of the peaks as the number of rows increases is very similar to the sharpening of the diffraction peaks from a diffraction grating as the number of slits increases.

X-RAY DIFFRACTION TECHNIQUES

Two types of targets are used in Bragg diffraction experiments, single crystals and powder targets.

Single Crystal

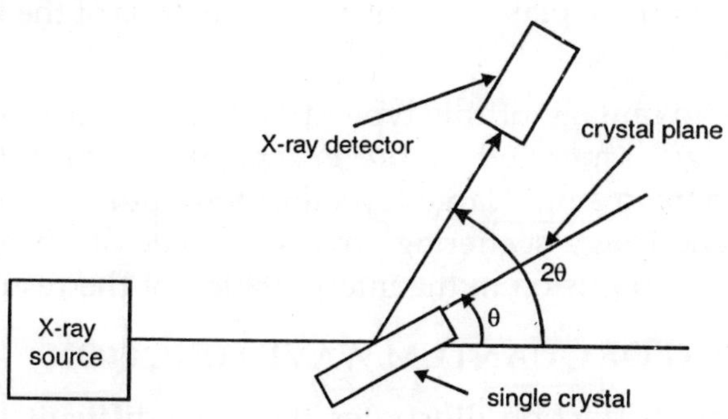

Fig. Setup for single crystal Bragg diffraction.

In a single crystal setup, an X-ray detector is mounted as shown in figure. A mechanical device keeps the detector oriented so that the angle of incidence equals the angle of reflection for the desired crystal plane. Peaks in the X-ray detection rate are sought as the angle θ is varied.

The advantage of this type of apparatus is that diffraction peaks from only the selected crystal plane are observed.

Powder Target

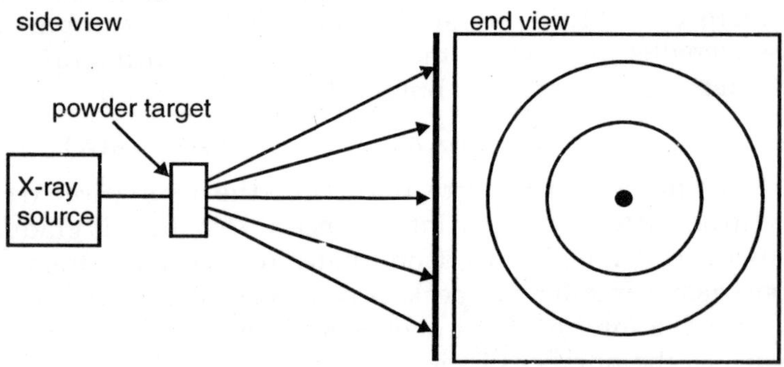

Fig. Setup for Powder Target Bragg Diffraction

The powder in a powder target is really a conglomeration of many tiny crystals randomly oriented. Thus, for each possible Bragg diffraction angle there are crystals oriented correctly for Bragg diffraction to take place. The detector is usually a photographic plate or an equivalent electronic device as illustrated in figure. For each Bragg diffraction angle one sees a ring on the plate concentric with the axis of the incident X-ray beam.

The advantage of this type of system is that no a priori knowledge is needed of the crystal plane orientations. Furthermore, a single large crystal is not required. However, all possible Bragg scattering angles are seen at once, which can lead to confusion in the interpretation of the results.

MEANING OF QUANTUM WAVE FUNCTION

Bragg diffraction illustrates the most difficult thing to understand about quantum mechanics, namely that particles can have wave-like properties and waves can have particle-like properties.

The variation of X-ray intensity with angle seen in a Bragg diffraction apparatus is very difficult to explain in any terms other than wave interference. Yet, X-rays are typically detected by a device such as a Geiger counter which produces a pulse of electricity for each X-ray particle, or *photon,* which hits it. Thus, X-rays sometimes act like particles and sometimes like waves.

Light isn't alone in having both particle and wave properties. Davisson and Germer and later G. P. Thomson (son of J. J. Thomson, the discoverer of the electron) showed that electrons also can act like waves. They did this by demonstrating that electrons undergo Bragg diffraction in crystals, much as X-rays do.

Most physicists (including Albert Einstein) have found quantum mechanics to be extremely bizarre, so if you feel the same way, you are in good company! However, there is a useful interpretation of quantum mechanics which at least allows us to get on with using it to solve problems, even though it may not satisfy our intuitive reservations about the theory.

The displacement of a quantum mechanical wave associated with a particle is usually called the wave function, ψ. Like light, it is not at all clear what ψ is a displacement *of,* but its use is straightforward. The absolute square of the wave function, $|\psi(x,t)|^2$, is proportional to the *probability* of finding the associated particle at position x and time t. The *absolute* square is taken because under many circumstances the wave function is actually *complex,* i. e., it has both real and imaginary parts. The reasons for this will be discussed later.

Due to the interpretation of the wave function, quantum mechanics is a probabilistic theory. It does not tell us with certainty what happens to a particular particle. Instead, it tells us, for instance, the probabilities for the particle to be detected in certain locations. If many experiments are done, with one particle per experiment, the numbers of experiments with particles being detected in the various possible locations are in proportion to the quantum mechanical probabilities.

SENSE AND NONSENSE IN QUANTUM MECHANICS

The essential mystery of quantum mechanics becomes clearer when discussed in the conceptually simpler context of two slit interference. If light and electrons can have both particle and wave properties, then one might ask through which of the two slits the particle passed. However, in physics a question simply doesn't make sense if it cannot be answered by experiment. Let us see if the above question makes sense.

One can indeed perform an experiment to determine which slit the X-ray photon or electron passes through in the two slit experiment. However, by the very act of making this measurement, the form of the associated wave is altered. In particular, since the absolute square of the wave displacement represents the probability of finding the particle, once the particle has definitely been found passing through one or the other of the slits, the wave function collapses into a very small wave packet located at the observed position of the particle. Thus, the wave displacement becomes zero at the slit it didn't pass through. However, the interference pattern results from the superposition of waves emanating from two slits. If no wave comes from one of the slits (because the wave displacement is zero there), then there can be no interference pattern.

We can now make the inverse argument. If there is an interference pattern, then we know that the wave displacement is non-zero at both slits. From the probability interpretation of the wave displacement, we conclude that we cannot say, even in principle, through which slit the particle passed. It is not just that we don't know the answer to this question; there is simply no experiment which can give us an answer without destroying the interference pattern. In other words, the question "Which slit did the particle pass through?" is a nonsensical question in the case where an interference pattern is actually produced.

The American physicist Richard Feynman noticed that the above behaviour can be interpreted as violating the normal laws of probability. These laws say that the probability of an event is the sum of the probabilities of alternate independent ways for that event to occur. For instance, the probability P_1 for a particle to reach point A on the detection screen of a two slit setup is just the probability P_2 for the particle to reach point A after going through slit 1, plus the probability for the particle to reach point A after going through slit 2. Thus, if $P_1 = P_2 = 0.1$, then the probability for the particle to reach point A irrespective of which slit it went through should be $P_{total} = P + P_2 = 0.2$. However, if point A happens to be a point of destructive interference, then we know that $P_{total} = 0$.

Feynman proposed that the above rule stating that alternate independent probabilities add, is simply incorrect. In its place Feynman asserted that probability amplitudes add instead, where the probability amplitude in this case is just the wave function associated with the particle. The probability is obtained by adding the alternate probability amplitudes together and taking the absolute square of the sum.

In the two slit case, let us call ψ_1 the amplitude for the particle to reach point A via slit 1, while ψ_2 is the amplitude for it to reach this point via slit 2. The total amplitude is therefore $\psi_{total} = \psi_1 + \psi_2$ and the overall probability for the particle to reach point A is $P_{total} = |\psi_1 + \psi_2|^2$. The wave functions ψ_1 and ψ_2 at point A are proportional respectively to $\sin(kL_1)$ and $\sin(kL_2)$, where k is the wavenumber L_1 and L_2 and are the respective distances from slit 1 and slit 2 to point A. It is easy to see that destructive interference occurs if the difference between L_1 and L_2 is half a wavelength, thus yielding $P_{total} = 0$.

The above treatment of two slits in terms of probability amplitudes yields results which are identical to the conventional treatment of the two slit problem for waves.

However, it is conceptually clearer in the case of quantum mechanics because we no longer need to ask whether "things" are really particles or waves. In particular, the particle/wave is replaced by just a particle plus the assertion that probability amplitudes rather than probabilities add. The laws of quantum mechanics then boil down to rules for how probability amplitudes behave. In the case of particles, the amplitudes behave like waves, which explains why the study of wave kinematics and dynamics is so important to our understanding of quantum mechanics.

MASS, MOMENTUM AND ENERGY

In this section we relate the classical ideas of mass, momentum and energy to what we have done so far. Historically, these connections were first made by Max Planck and Louis de Broglie with help from Albert Einstein. Bragg diffraction by electrons is invoked as an experimental test of the Planck and de Broglie relations.

Technically, we don't need the ideas of mass, momentum, and energy to do physics - the notions of wavenumber, frequency, and group velocity are sufficient to describe and explain all observed phenomena. However, mass, momentum, and energy are so firmly embedded in physics that one couldn't talk to other physicists without an understanding of these quantities.

Planck, Einstein, and de Broglie

Max Planck was the first to make a connection between the energy of a particle and the frequency of the associated wave. Planck developed his formula in his successful attempt to explain the emission and absorption of electromagnetic radiation from warm material surfaces. It suffices here to convey Planck's formula, which was originally intended to apply only to parcels of electromagnetic radiation, or *photons*:

$E = hf = \hbar w$ (Planck relation),

where E is the *energy* of the photon, f is the rotational frequency of the associated light wave, and w is its angular frequency. The constant h = 6.63 × 10–34 kgm^{2}s^{-1} is called *Planck's constant*. The related constant $\hbar = h/2\pi = 1.06 \times 10^{-34}$ kgm^{2}s^{-1} is also referred to as Planck's constant, but to avoid confusion with the original constant, we will generally refer to it as "h bar".

Notice that a new physical dimension has appeared, namely mass, with the unit kilogram, abbreviated "kg". The physical meaning of mass is much like our intuitive understanding of the concept, i. e., as a measure of the resistance of an object to its velocity being changed. The precise scientific meaning will emerge shortly.

Einstein showed that Planck's idea could be used to explain the emission of electrons which occurs when light impinges on the surface of a metal. This emission, which is called the photoelectric effect, can only occur when electrons are supplied with a certain minimum energy E_B required to break them loose from the metal. Experiment shows that this emission occurs only when the frequency of the light exceeds a certain minimum value. This value turns out to equal $W_{\text{min}} = E_B / \hbar$, which suggests that electrons gain energy by absorbing a single photon. If the photon energy, $\hbar w$, exceeds E_B, then electrons are emitted, otherwise they are not. It is much more difficult to explain the photoelectric effect from the classical theory of light.

Louis de Broglie proposed that Planck's energy-frequency relationship be extended to all kinds of particles. In addition he hypothesized that the momentum Π of the particle and the wave vector k of the corresponding wave were similarly related:

$\Pi = \hbar k$ (de Broglie relation)

Note that this can also be written in scalar form in terms of the wavelength as $\Pi = h/\lambda$.

De Broglie's hypothesis was inspired by the fact that wave frequency and wavenumber are components of the same four-vector according to the theory of relativity, and are therefore closely related to each other. Thus, if the energy of a particle is related to the frequency of the corresponding wave, then there ought to be some similar quantity which is correspondingly related to the wavenumber.

It turns out that the momentum is the appropriate quantity. The physical meaning of momentum will become clear as we proceed.

We will also find that the rest frequency, μ, of a particle is related to its mass, m:

$$E_{rest} = mc^2 = \hbar\mu$$

The quantity E_{rest} is called the rest energy of the particle.

From our perspective, energy, momentum, and rest energy are just scaled versions of frequency, wave vector, and rest frequency, with a scaling factor $\hbar$. We can therefore define a *four-momentum* as a scaled version of the wave four-vector:

$$\underline{\Pi} = \hbar\underline{k}.$$

The spacelike component of $\underline{\Pi}$ is just Π, while the timelike part is E/c.

Planck, Einstein, and de Broglie had extensive backgrounds in classical mechanics, in which the concepts of energy, momentum, and mass have precise meaning. In this text we do not presuppose such a background. Perhaps the best strategy at this point is to think of these quantities as scaled versions of frequency, wavenumber, and rest frequency, where the scale factor is $\hbar$. The significance of these quantities to classical mechanics will emerge bit by bit.

WAVE AND PARTICLE QUANTITIES

Let us now recapitulate what we know about relativistic waves, and how this knowledge translates into knowledge about the mass, energy, and momentum of particles. In the following equations, the left form is expressed in wave terms, i. e., in terms of frequency, wavenumber, and rest frequency. The right form is the identical equation expressed in terms of energy, momentum, and mass. Since the latter variables are just scaled forms of the former, the two forms of each equation are equivalent.

We begin with the dispersion relation for relativistic waves:

$$w^2 = k^2c^2 + \mu^2 \qquad E^2 = \Pi^2c^2 + m^2c^4$$

Calculation of the group velocity, $u_g = dw/dk$, from the dispersion relation yields

$$u_g = \frac{c^{2k}}{w} \qquad u_g = \frac{c^2\Pi}{E}$$

These two sets of equations represent what we know about relativistic waves, and what this knowledge tells us about the relationships between the mass, energy, and momentum of relativistic particles. When in doubt, refer back to these equations, as they work in all cases, including for particles with zero mass!

It is useful to turn equations around so as to express the frequency as a function of rest frequency and group velocity,

$$w = \frac{\mu}{\left(1 - u_g^2/c^2\right)^{1/2}} \qquad E = \frac{mc^2}{\left(1 - u_g^2/c^2\right)^{1/2}}$$

and the wavenumber as a similar function of these quantities:

$$k = \frac{\mu u_g/c^2}{\left(1 - u_g^2/c^2\right)^{1/2}} \qquad \Pi = \frac{mu_g}{\left(1 - u_g^2/c^2\right)^{1/2}}$$

Note that equations work only for particles with non-zero mass! For zero mass particles you need to use equations with $m = 0$ and $\mu = 0$.

The quantity $w - \mu$ indicates how much the frequency exceeds the rest frequency. Notice that if $w = \mu$, then from equation $k = 0$. Thus, positive values of $k = w - \mu$

indicate $|k| > 0$, which means that the particle is moving according to equation. Let us call k the kinetic frequency:

$$k = \left[\frac{1}{\left(1 - u_g^2 / c^2\right)^{1/2}} - 1\right]\mu \qquad k = \left[\frac{1}{\left(1 - u_g^2 / c^2\right)^{1/2}} - 1\right]mc^2$$

We call K the kinetic energy for similar reasons. Again, equation only works for particles with non-zero mass. For zero mass particles the kinetic energy equals the total energy.

Non-Relativistic Limits

When the mass is non-zero and the group velocity is much less than the speed of light, it is useful to compute approximate forms of the above equations valid in this limit. Using the approximation $(1+\in)^x \approx 1 + x\in$, we find that the dispersion relation becomes

$$w = \mu + \frac{k^2c^2}{2\mu} \qquad E = mc^2 + \frac{\Pi^2}{2m}$$

and the group velocity equation takes the approximate form

$$u_g = \frac{c^2 k}{\mu} \qquad u_g = \frac{\Pi}{m}$$

The non-relativistic limits for equations become

$$w = \mu + \frac{\mu u_g^2}{2c^2} \qquad E = mc^2 + \frac{m u_g^2}{2}$$

and

$$k = \mu u_g / c^2 \quad \Pi = m u_g$$

while the approximate kinetic energy equation is

$$k = \frac{\mu u_g^2}{2c^2} \qquad K = \frac{\mu u_g^2}{2}$$

Just a reminder – the equations in this section are not valid for massless particles.

HEISENBERG UNCERTAINTY PRINCIPLE

Classically, we consider the location of a particle to be a knowable piece of information. In quantum mechanics the position of a particle is well known if the wave packet representing it is small in size. However, quantum mechanics imposes a price on accurately knowing the position of a particle in terms of the future predictability of its position. This is because a small wave packet, which corresponds to accurate knowledge of the corresponding particle's position, implies the superposition of plane waves corresponding to a broad distribution of wavenumbers. This translates into a large uncertainty in the wavenumber, and hence the momentum of the particle. In contrast, a broad wave packet corresponds to a narrower distribution of wavenumbers, and correspondingly less uncertainty in the momentum.

The longitudinal (along the direction of motion) and transverse (normal to the direction of motion) dimensions of a wave packet, Δx_L and Δx_T, can be related to the spread of longitudinal and transverse wavenumbers, ΔK_L and ΔK_T :

$$\Delta K_L \Delta x_L \approx 1$$

$$\Delta K_T \Delta x_T \approx 1$$

We have omitted numerical constants which are order unity in these approximate relations so as to show their essential similarity.

The above equations can be interpreted in the following way. Since the absolute square of the wave function represents the probability of finding a particle, Δx_L and Δx_T represent the uncertainty in the particle's position. Similarly, Δk_L and Δk_T represent the uncertainty in the particle's longitudinal and transverse wave vector components.

This latter uncertainty leads to uncertainty in the particle's future evolution — larger or smaller longitudinal k results respectively in larger or smaller particle speed, while uncertainty in the transverse wavenumber results in uncertainty in the particle's direction of motion. Thus uncertainties in any component of k result in uncertainties in the corresponding component of the particle's velocity, and hence in its future position.

The equations show that uncertainty in the present and future positions of a particle are complimentary. If the present position is accurately known due to the small size of the associated wave packet, then the future position is not very predictable, because the wave packet disperses rapidly. On the other hand, a broad-scale initial wave packet means that the present position is poorly known, but the uncertainty in position, poor as it is, doesn't rapidly increase with time, since the wave packet has a small uncertainty in wave vector and thus disperses slowly. This is a statement of the Heisenberg uncertainty principle.

The uncertainty principle also applies between frequency and time:

$$\Delta w \Delta t \approx 1$$

This shows up in the beat frequency equation $1/T_{beat} = \Delta f = \Delta w / 2\pi$. The beat period may be thought of as the size of a "wave packet in time". The beat frequency equation T_{beat} may be rewritten as $\Delta w T_{beat} = 2\pi$, which is the same as equation if the factor of 2π is ignored and is T_{beat} identified with Δt.

The above forms of the uncertainty principle are not relativistically invariant. A useful invariant form may be obtained by transforming to the coordinate system in which a particle is stationary. In this reference frame the time t becomes the proper time τ associated with the particle. Furthermore, the frequency w becomes the rest frequency μ. The uncertainty principle thus becomes

$$\Delta \mu \Delta \tau \approx 1$$

in this reference frame. However, since $\Delta \mu$ and $\Delta \tau$ are relativistic invariants, this expression of the uncertainty principle is valid in *any* reference frame.

It is more common to express the uncertainty principle in terms of the mass, momentum, and energy by multiplying equations by $\hbar$. Lumping the momentum equations, we find

$$\Delta\Pi\Delta x \approx \hbar$$

$$\Delta E\Delta x \approx \hbar$$

and

$$\Delta\left(mc^2\right)\Delta\tau \approx \hbar$$

Classical mechanics is the realm of quantum mechanics in which the dimensions of the system of interest are much larger than the wavelengths of the waves corresponding to the particles constituting the system. In this case the uncertainties induced by the uncertainty principle are unimportant. This limit is analogous to the geometrical optics limit for light. Thus, we can say that classical mechanics is the geometrical optics limit of quantum mechanics.

Chapter 5

Quantum Dynamics

In quantum mechanics, particles are represented by matter waves, with the absolute square of the wave displacement yielding the probability of finding the particle. The behaviour of particles thus follows from the reflection, refraction, diffraction, and interference of the associated waves. The connection with Newtonian dynamics comes from tracing the trajectories of matter wave packets. Changes in the speed and direction of motion of these packets correspond to the accelerations of classical mechanics. When wavelengths are small compared to the natural length scale of the problem at hand, the wave packets can be made small, thus pinpointing the position of the associated particle, without generating excessive uncertainty in the particle's momentum. This is the geometrical optics limit of quantum mechanics.

ONE-DIMENSIONAL NON-RELATIVISTIC CASE

Louis de Broglie made an analogy between matter waves and light waves, pointing out that wave packets of light change their velocity as the result of spatial variations in the index of refraction of the medium in which they are travelling. This behaviour comes about because the dispersion relation for light traveling through a medium with index of refraction n is $w = kc/n$, so that the group velocity, $u_g = dw/dk = c/n$. Thus, when n increases, u_g decreases, and vice versa.

The problem for matter waves is to determine how analogous modifications can be made to the free particle dispersion relation so as to produce accelerations of wave packets consistent with Newtonian dynamics in the geometrical optics limit. In this section we make a simple guess as to how to modify the free particle dispersion relation for matter waves, limiting consideration initially to the one-dimensional, non-relativistic case. As in many situations in physics, the simple guess turns out to be correct! This leads to a connection between wave dynamics and Newtonian mechanics.

The dispersion relation for free matter waves is w = $\left(k^2c^2+\mu^2\right)^{1/2}$. In the non-relativistic limit $k^2c^2 \ll \mu^2$ and this equation can be approximated as

$$w = \mu\left(1+k^2c^2/\mu^2\right)^{1/2} \approx \mu+k^2c^2/(2\mu)$$

Let us modify this equation by adding a term S' which depends on x, and which plays a role analogous to the index of refraction for light:

$$w = S'(x)+\mu+k^2c^2/(2\mu)=S(x)+k^2c^2/(2\mu)$$

The rest frequency has been made to disappear on the right side of the above equation by defining $S = S'+\mu$. This is done to simplify the notation.

Let us now imagine that all parts of the wave governed by this dispersion relation oscillate in phase. The only way this can happen is if w is constant, i. e., it takes on the same value in all parts of the wave. It turns out we can do this if S is not a function of time.

If w is constant, the only way S can vary with x in equation is if the wavenumber varies in a compensating way. Thus, constant frequency and spatially varying S together imply that $k = k(x)$. Solving equation for k yields

$$k(x) = \left[\frac{2\mu[w - S(x)]}{c^2}\right]^{1/2}$$

Since w is constant, the wavenumber becomes smaller and the wavelength larger as the wave moves into a region of increased S.

In the geometrical optics limit, we assume that S doesn't change much over one wavelength so that the wave remains reasonably sinusoidal in shape with approximately constant wavenumber over a few wavelengths. However, over distances of many wavelengths the wavenumber and amplitude of the wave are allowed to vary considerably.

As yet we have no idea what causes S or where it comes from. For now we simply explore the consequences of its presence, especially in the geometric optics limit in which quantum dynamics gives way to Newtonian dynamics.

The group velocity calculated from the dispersion relation given by equation is

$$u_g = \frac{dw}{dk} = \frac{kc^2}{\mu} = \left(\frac{2c^2(w - S)}{\mu}\right)^{1/2}$$

where k is eliminated in the last step with the help of equation. The resulting equation tells us how the group velocity varies as a matter wave traverses a region of slowly varying S. Thus, as S increases, u_g decreases and vice versa.

We can now calculate the acceleration of a wave packet resulting from the spatial variation in S. We assume that $x(t)$ represents the position of the wave packet, so that $u_g = dx/dt$. Using the chain rule $du_g/dt = (du_g/dx)(dx/dt)$ = $(du_g/dx)^{\mu_g}$, we find

$$a = \frac{du_g}{dt} = \frac{du_g}{dx}\mu_g = \frac{du_g^2/2}{dx} = -\frac{c^2}{\mu}\frac{dS}{dx} = -\frac{\hbar}{m}\frac{dS}{dx}$$

The group velocity is eliminated in favour of *S* by squaring equation and substituting the result into equation.

Using Newton's second law to infer the force from the acceleration, we find

$$F = ma = -\frac{dU}{dx} \text{ (force from potential energy)}$$

where we have defined $U \equiv \hbar s$. The quantity U is called the potential energy. We have thus established a relationship between quantum mechanical and Newtonian dynamics, in that $U \equiv \hbar s$ dictates the form of the force in Newton's second law, while *S* governs the refraction of matter waves. A force which equals minus the derivative of some potential energy U(x) is called conservative. Certain forces such as friction are not conservative. At present we will deal mainly with conservative forces.

Potential Energy

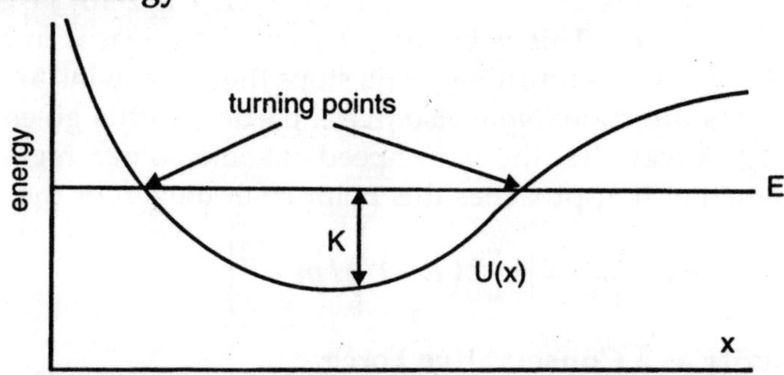

Fig: Example of spatially variable potential energy *U*(*x*) with fixed total energy E. The kinetic energyK = E – U is zero where the *E* and *U* lines cross. These points are called turning points.

Recall that the momentum is related to the group velocity by

$$\Pi = mu_g \text{ (momentum)}$$

in the nonrelativistic case. Multiplying equation by $\hbar$ and eliminating k and μ in favour of Π and m results in

$$E = U + \Pi^2 / (2m) = U + K \text{ (total energy)}$$

where we have used $U = \hbar S$ and $\hbar \mu = mc^2$, and where

$$K = \Pi^2 / (2m) = mu_g^2 / 2 \text{ (kinetic energy)}$$

is the nonrelativistic kinetic energy.

The energy *E*, which we now refer to for clarity as the total energy, is constant since it is proportional to the (constant) frequency *w*. We say that the total energy is conserved. Thus, increases in the potential energy coincide with decreases in the kinetic energy and vice versa, as is illustrated in figure. In Newtonian dynamics the kinetic energy cannot be negative, since it is the product of half the mass and the square of the velocity, both of which are positive. Thus, a particle with total energy *E* and potential energy *U* is forbidden to venture into regions in which the kinetic energy $K = E - U$ is negative.

The points at which the kinetic energy is zero are called turning points. This is because a particle decreases in speed as it approaches a turning point, stops there for an instant, and reverses direction. Note also that a particle with a given total energy always has the same speed at some point *x* regardless of whether it approaches this point from the left or the right:

$$speed = |u_g| = \left| \pm \left[2(E - U) / m \right]^{1/2} \right|$$

Gravity as a Conservative Force

An example of a conservative force is gravity. An object of mass *m* near the surface of the earth has the gravitational potential energy

$$U = mgz \text{ (gravity near earth's surface)}$$

where z is the height of the object and g = 9.8ms^{-2} is the local value of the gravitational field near the earth's surface. Notice that the gravitational potential energy increases upward. The speed of the object in this case is $|u_g| = \left[2(E - mgz) / m \right]^{1/2}$.

If $\left|u_g\right|$ is known to equal the constant value u_0 at elevation z = 0, then equations tell us that $u_0 = (2E/m)^{1/2}$ and $\left|u_g\right| = \left(u_o^2 - 2gz\right)^{1/2}$.

There are certain types of questions which energy conservation cannot directly answer. For instance if an object is released at elevation h with zero velocity at $t = 0$, at what time will it reach z = 0 under the influence of gravity? In such cases it is often easiest to return to Newton's second law. Since the force on the object is in F = –dU/dz = –mg this case, we find that the acceleration is a = F/m = –mg/m = –g. However, a = du/dt = d^2z/dt^2, so

$$u = -gt + C_1 \quad z = -gt^2/2 + C_1t + C_2 \text{ (constant gravity).}$$

where C_1 and C_2 are constants to be determined by the initial conditions. These results can be verified by differentiating to see if the original acceleration is recovered. Since u = 0 and z = h at t = 0, we have $C_1 = 0$ and $C_2 = h$. With these results it is easy to show that the object reaches z = 0 when $t = (2h/g)^{1/2}$.

Work and Power

When a force is exerted on an object, energy is transferred to the object. The amount of energy transferred is called the work done on the object. However, energy is only transferred if the object moves. The work W done is

$$W = F\Delta x$$

where the distance moved by the object is Δx and the force exerted on it is F. Notice that work can either be positive or negative. The work is positive if the object being acted upon moves in the same direction as the force, with negative work occurring if the object moves opposite to the force.

Equation assumes that the force remains constant over the full displacement Δx. If it is not, then it is necessary to break up the displacement into a number of smaller displacements, over each of which the force can be assumed to be constant. The total work is then the sum of the works associated with each small displacement.

If more than one force acts on an object, the works due to the different forces each add or subtract energy, depending on whether they are positive or negative. The total work is the sum of these individual works.

There are two special cases in which the work done on an object is related to other quantities. If F is the total force acting on the object, then $W = F\Delta x = ma\Delta x$ by Newton's second law. However, $a = dv/dt$ where is the velocity of the object, and $\Delta x = (\Delta x/\Delta t)\Delta t \approx v\Delta t$, where Δt is the time required by the object to move through distance Δx. The approximation becomes exact when Δx and Δt become very small. Putting all of this together results in

$$W_{total} = m\frac{dv}{dt}v\Delta t = \frac{d}{dt}\left(\frac{mv^2}{2}\right)\Delta t = \Delta K \ (total\ work),$$

where K is the kinetic energy of the object. Thus, when F is the only force, $W = W_{total}$ is the total work on the object, and this equals the change in kinetic energy of the object. This is called the work-energy theorem, and it demonstrates that work really is a transfer of energy to an object.

The other special case occurs when the force is conservative, but is not necessarily the total force acting on the object. In this case

$$W_{cons} = -\frac{dU}{dx}\Delta x = -\Delta U \ \text{(conservative force)},$$

where ΔU is the change in the potential energy of the object associated with the force of interest.

The power associated with a force is simply the amount of work done by the force divided by the time interval Δt over which it is done. It is therefore the energy per unit time transferred to the object by the force of interest. From equation we see that the power is

$$P = \frac{F\Delta x}{\Delta t} = Fv \text{ (Power)},$$

where is the velocity at which the object is moving. The total power is just the sum of the powers associated with each force. It equals the time rate of change of kinetic energy of the object:

$$P_{total} = \frac{W_{total}}{\Delta t} = \frac{dK}{dt} \text{ (total power)}.$$

Math Tutorial - Partial Derivatives

In order to understand the generalization of Newtonian mechanics to two and three dimensions, we first need to understand a new type of derivative called the partial derivative.

To motivate the discussion, let us use the chain rule to take the ordinary derivative of $f(x, y) = x^2 + y^2$ where y is some function of x, say $y = x^3 + x$:

$$\frac{df}{dx} = 2x + 2y\frac{dy}{dx} = 2x + 2y.(3x^2 + 1).$$

At times we may wish to take the derivative of f with respect to while ignoring the possible dependence of y on x. A derivative of this type has special notation — the "d" is replaced by "∂" and the derivative is called a partial derivative. Thus, the partial derivative of with f respect to x is

$$\frac{\partial f}{\partial x} = 2x.$$

Notice that the partial derivative is actually simpler to evaluate than an ordinary derivative, because only the explicit dependence of the function on the differentiating variable need be considered — all other variables are taken to be constant.

Motion in Two and Three Dimensions

When a matter wave moves through a region of variable potential energy in one dimension, only the wavenumber changes. In two or three dimensions the wave vector can change in both direction and magnitude. This complicates the calculation of particle movement. However, we already have an example of how to handle this situation, namely, the refraction of light. In that case Snell's law tells us how the direction of the wave vector changes, while the dispersion relation combined with the constancy of the frequency gives us information about the change in the magnitude of the wave vector. For matter waves a similar procedure works, though the details are different, because we seek the consequences of a change in potential energy rather than a change in the index of refraction.

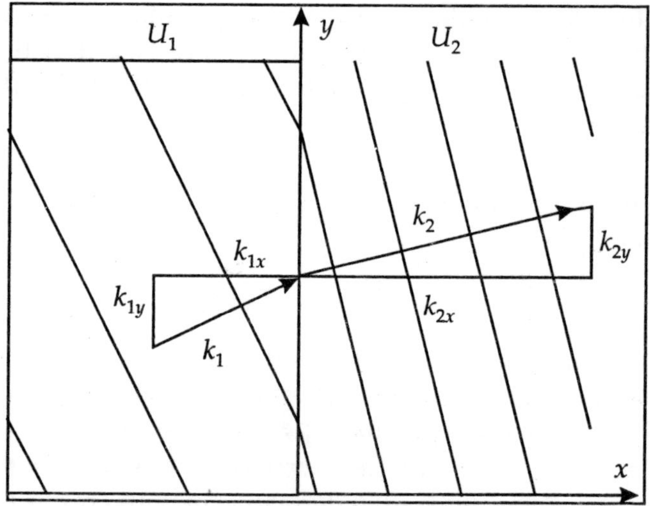

Fig. Refraction of a matter wave by a discontinuity in potential energy. The component of the wave vector parallel to the discontinuity, k_y, doesn't change, so $k_{1y} = k_{2y}$.

Figure illustrates the refraction of matter waves at a discontinuity in the potential energy. Let us suppose that the discontinuity occurs at $x = 0$. If the matter wave to the left of the discontinuity is $\psi_1 = \sin(k_{1x}x + k_{1y}y - \omega_1 t)$ and to the right is $\psi_2 = \sin(k_{2x}x + k_{2y}y - \omega_2 t)$, then the wavefronts of the waves will match across the discontinuity for all time only if $\omega_1 = \omega_2 \equiv \omega$ and $k_{1y} = k_{2y} \equiv k_y$. We are already familiar with the first condition from the one-dimensional problem, so the only new ingredient is the constancy of the component of the y wave vector.

In two dimensions the momentum is a vector: $\prod = mu$ when $|u| \ll c$, where u is the particle velocity. Furthermore, the kinetic energy is $K = m|u|^2/2 = |\pi|^2/(2m) = (\Pi_x^2 + \Pi_y^2)/$ (2m). The relationship between kinetic, potential, and total energy is unchanged from the one-dimensional case, so we have

$$E = U + (\Pi_x^2 + \Pi_y^2)/(2m) = \text{constant.}$$

The de Broglie relationship tells us that $\Pi = \hbar k$, so the constancy of k_y across the discontinuity in U tells us that Π_y = constant. there.

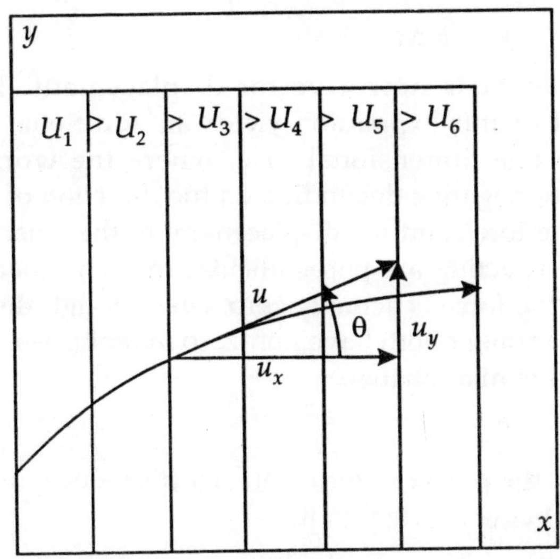

Fig. Trajectory of a wave packet through a variable potential energy, $U(x)$, which decreases to the right.

Let us now approximate a continuously variable $U(x)$ by a series of steps of constant U oriented normal to the x-axis. The above analysis can be applied at the jumps or discontinuities in U between steps, as illustrated in figure, with the result that equations are valid across all discontinuities. If we now let the step width go to zero, these equations then become valid for U continuously variable in x.

Using the procedure which we invoked before, we find the force components associated with $U(x)$ in the x and y directions to be $F_x = - dU/dx$ and $F_y = 0$. This generalizes to

$$F = -\left(\frac{\partial U}{\partial x}, \frac{\partial U}{\partial y}, \frac{\partial U}{\partial z}\right) \text{ (3-D conservative force)}$$

in the three-dimensional case where the orientation of constant U surfaces is arbitrary. It is also valid when $U(x, y, z)$ is not limited to a simple ramp form, but takes on a completely arbitrary structure.

The definitions of work and power are slightly different in two and three dimensions. In particular, work is defined

$$W = F.\Delta x$$

where Δx is now a vector displacement. The vector character of this expression yields an additional possibility over the one dimensional case, where the work is either positive or negative depending on the direction of Δx relative to F. If the force and the displacement of the object on which the force is acting are perpendicular to each other, the work done by the force is actually zero, even though the force and the displacement both have non-zero magnitudes. The power exhibits a similar change:

$$P = F.u.$$

Thus the power is zero if an object's velocity is normal to the force being exerted on it.

Energy conservation by itself is somewhat less useful for solving problems in two and three dimensions than it is in one dimension. This is because knowing the kinetic energy at some point tells us only the magnitude of the velocity, not its direction.

If conservation of energy fails to give us the information we need, then we must revert to Newton's second law, as we did in the one-dimensional case. For instance, if an object of mass has initial velocity $u_0 = (u_0, 0)$ at location $(x, z) = (0, h)$ and has the gravitational potential energy $U = mgz$, then the force on the object is $F = (0, -mg)$. The acceleration is therefore $a = F/m = (0, -g)$. Since $a = du/dt = d^2x/dt^2$ where $x = (x, z)$ is the object's position, we find that

$$u = (C_1, -gt + C_2)\ x = (C_1t + C_3, -gt^2/2 + C_2t + C_4),$$

where C_1, C_2, C_3, and C_4 are constants to be evaluated so that the solution reduces to the initial conditions at $t = 0$. The specified initial conditions tell us that $C_1 = u_0$, $C_2 = 0$, $C_3 = 0$, and $C_4 = h$ in this case. From these results we can infer the position and velocity of the object at any time.

KINETIC AND POTENTIAL MOMENTUM

If you have previously taken a physics course then you have probably noticed that a rather odd symbol is used for momentum, namely Π, rather than the more commonly employed P. The reason for this peculiar usage is that there are actually two kinds of momentum, kinetic momentum and total momentum, just as there are two kinds of energy, kinetic and total. The symbol Π represents total momentum while P represents kinetic momentum. In non-relativistic mechanics we don't need to distinguish between the two quantities, as they are generally equal to each other. However, we will find later in the course that it is crucial to make this distinction in the relativistic case. As a general rule, the total momentum is related to a particle's wave vector via the de Broglie relation, $\Pi = \hbar k$, while the kinetic momentum is related to a particle's velocity, $P = mu/(1 - u^2/c^2)^{1/2}$.

Chapter 6

Relativistic Electronic Structure

Relativistic effects have been found to play a crucial role in the description of the electronic structure of heavy elements. *Ab initio* quantum mechanical methods which are based upon many-electron extensions of Dirac's relativistic single-particle wave equation have proven to be extremely effective in properly describing the electronic structure of molecules which contain heavy atoms, though these methods are prohibitively expensive for most systems of chemical interest. These notes begin with a review of Dirac's equation for a single particle, and some many-electron extensions of this equation. Next, many of the most widely used relativistic quantum chemical techniques are discussed. Finally, the major effects of special relativity on atomic electronic structure are explored and quantified, and the consequences of these effects on molecular structure and chemical behaviour are discussed.

The Dirac Equation

One of the first attempts at a relativistic quantum mechanical wave equation was the Klein-Gordon (KG) equation. Though it is theoretically sound from the perspective that it is consistent with both classical quantum mechanics and the special theory of relativity, it has several unsavory features which keep it from being a very powerful tool in relativistic quantum mechanics. Dirac later developed his own relativistic wave equation which did not have some of the shortcomings of the Klein-Gordon equation. Dirac's formulation also led to the discovery that some of the supposed "errors" of the KG equation were actually illustrating some new physics. The Dirac equation is only rigorous for a one particle system, but has been used as a starting point for a number of approximate many electron methods.

KLEIN-GORDON EQUATION

The special-relativistic expression for the kinetic energy may be used to form the classical free particle Hamiltonian

$$E = c(m^2c^2 + p^2)^{1/2}.$$

The analogous quantum mechanical expression may be constructed by replacing the classical momentum, p, with its quantum mechanical operator, $\hat{p}$ which yields the free particle wave equation

$$\left(i\hbar\frac{\partial}{\partial t}\right)\psi = [c(m^2c^2 + \hat{p}^2)^{1/2}]\psi$$

This equation, however, does not satisfy some of the conditions required by special relativity: The wave equation is not invariant to a Lorentz transform, and the square root term introduces ambiguity. The Klein-Gordon equation rectifies both of these problems simply by taking the square of the original energy expression and extending the result to a quantum mechanical wave equation:

$$E^2\psi = \left(i\hbar\frac{\partial}{\partial t}\right)^2 \psi = (m^2c^4 + \hat{p}^2c^2)\psi$$

The resulting wave equation is Lorentz covariant and well defined, but suffers from other problems. Negative energy solutions to this equation are possible, which do not have a readily obvious explanation, and the probability density, $\psi^*\psi$, fluctuates with time as does its integral over all space.

The foibles of the Klein-Gordon equation may make it a poor equation for the electron, but those weaknesses helped to point Dirac in the right direction and to develop a single particle equation which successfully surmounted all these problems.

Dirac's Free Particle Equation

In order for a wave equation to satisfy the special relativistic requirement of Lorentz covariance, derivatives in space and time must all appear in the same order. The K-G equation illustrates that an expression which satisfies this condition but is non-linear in the space and time derivatives gives rise to anomalous results. Dirac set out to find an equation which was first order in space and time derivatives. The result of his efforts, the Dirac equation, is difficult to motivate and far more complicated that the non-relativistic analog. However, Dirac's wave equation for a single particle satisfies all the requirements of special relativity and quantum mechanics, and is able to predict the properties of one particle systems with remarkable accuracy.

Dirac's equation for an electron in field-free space is given by

$$(\hat{p}_0 - \alpha.\hat{p} - \beta mc)\psi = 0$$

where

$$\hat{p}_0 = i\frac{\hbar}{c}\frac{\partial}{\partial t}$$

and $\hat{p}$ is simply the three component momentum. In order to determine the nature of the three components of α and β, it is useful to compare the modified equation

$$(\hat{p}_0 - \alpha.\hat{p} - \beta mc)^*(\hat{p}_0 - \alpha.\hat{p} - \beta mc)\psi = 0$$

to the Klein-Gordon equation. Equations are equivalent if we enforce the conditions

$$[\alpha_i, \alpha_j] = \alpha_i\alpha_j + \alpha_j\alpha_i = \delta_{ij}$$

where α_i represents β for i = 0 and α_x, α_y, and α_z for i = 1 – 3, respectively. In order for this set of four objects to fulfill these anti-commutation relations, each must be, minimally, four-dimensional.

One set of matrices which obey similar anti-commutation relations are the Pauli spin matrices: $\sigma_{x,y,z}$. These are 2 × 2 matrices, however, and there are only three of them, so they are not useful in their usual form. The anti-commutation conditions given in equation are satisfied if the α_i's are defined as

$$\alpha_0 = \begin{pmatrix} I_{2\times 2} & 0 \\ 0 & -I_{2\times 2} \end{pmatrix},$$

$$\alpha_i = \begin{pmatrix} 0 & \sigma_i \\ \sigma_0 & 0 \end{pmatrix} (i = 1 - 3)$$

where $I_{2\times 2}$ is a two by two identity matrix and σ_i are the Pauli spin matrices. Though these four by four matrices do not represent the only set of matrices which satisfy the anti-commutation relations, any set which does may only differ by a similarity transform from this set.

Because the Dirac equation contains operators represented by four dimensional matrices, the solutions $\{\psi\}$ must be represented by a four component vector

$$\psi = \begin{pmatrix} \psi_1 \\ \psi_2 \\ \psi_3 \\ \psi_4 \end{pmatrix}$$

The interpretation of the components of this four-vector is not readily evident. Expressing the Dirac equation in full matrix form provides some elucidation

$$\begin{pmatrix} (\hat{p}_0 - mc) & 0 & -\hat{p}_z & -(\hat{p}_x - i\hat{p}_y) \\ 0 & (\hat{p}_0 - mc) & -(\hat{p}_x + i\hat{p}_y) & \hat{p}_z \\ -\hat{p}_z & -(\hat{p}_x - i\hat{p}_y) & (\hat{p}_0 + mc) & 0 \\ -(\hat{p}_x + i\hat{p}_y) & \hat{p}_z & 0 & (\hat{p}_0 + mc) \end{pmatrix} \begin{pmatrix} \psi_1 \\ \psi_2 \\ \psi_3 \\ \psi_4 \end{pmatrix} = 0$$

The non-relativistic electronic wave-function has two components corresponding to the α and β components of spin angular momentum. In the non-relativistic limit, $\hat{p}_0$

approaches mc, and the terms which couple ψ_1 with ψ_2 drop out. What remains are four eigenvector equations, with approximate eigenvalues of + m_0c^2 for ψ_1 and ψ_2, and $-m_0c^2$ for ψ_3 and ψ_4. ψ_1 and ψ_2, then, may be interpreted as the α and β components of positive energy, electron-like solutions, but the solutions which are dominated by ψ_3 and ψ_4 do not posses a readily evident interpretation.

The existence of these negative energy solutions troubled Dirac. He felt that it was a "great blemish" on his theory that these apparently unphysical solutions remained. Over the next few years, however, several experimental investigations, most notably those of Carl Anderson, gave evidence for the existence of positrons, particles with the mass of an electron but a positive charge. Experimental confirmation of the existence of these negative energy solutions was a great triumph for Dirac's theory.

Hydrogenic Solutions of Dirac's Equation

In its present form, the Dirac equation cannot even address the hydrogen atom, since it has no provision for an external potential. If we introduce scalar and vector potentials, ϕ and A, to the momenta of the field free Dirac equation

$$p_0 \rightarrow p_0 - \frac{e\phi}{c}$$

$$\hat{p} \rightarrow \hat{p} - \frac{A}{c} = \hat{\pi}$$

then the Dirac equation becomes

$$\left[p_0 - \frac{\phi}{c} - \alpha.\hat{\pi} - \beta mc\right]\psi = 0$$

If the potential terms are not explicitly time dependent, we may make the rearrangement

$$\left[\beta mc^2 + c\alpha.\hat{\pi} + e\phi\right]\psi = E\psi$$

The left hand side defines the Dirac Hamiltonian for a single particle in a time-independent electro-magnetic field.

In the absence of a magnetic field the Dirac Hamiltonian, $\hat{H}_D$, may be expressed in full matrix form as

$$\begin{pmatrix} (e\phi + mc^2) & 0 & c\hat{p}_z & c(\hat{p}_x - i\hat{p}_y) \\ 0 & (e\phi + mc^2) & c(\hat{p}_x + i\hat{p}_y) & -c\hat{p}_z \\ c\hat{p}_z & c(\hat{p}_x - i\hat{p}_y) & (e\phi - mc^2) & 0 \\ c(\hat{p}_x + i\hat{p}_y) & -c\hat{p}_z & 0 & (e\phi - mc^2) \end{pmatrix}$$

This Hamiltonian is now ready to take on the task of an electron in the field of a nucleus.

Because of the spherical symmetry of the nuclear potential, the Schrödinger equation for the hydrogenic atom may be simplified via separation of variables into a radial equation and two angular equations. The angular solutions are given by the spherical harmonics, $Y_{1m_1}(\theta, \phi)$, and the radial solutions by

$$R_{nI}(r) = N_{nI} e^{-\frac{r}{na_0}} \left(\frac{2r}{na_0}\right)^I L_{n+I}^{2I+1}\left(\frac{2r}{na_0}\right)$$

where $L_{n+I}^{2I+1}(x)$ are the associated Laguerre polynomials, and *n*, *l*, and m_I are the principal quantum number, orbital angular momentum quantum number, and the z-projection of l, respectively.

The hydrogenic solutions to the Dirac equations may also be achieved analytically, though the resulting eigenfunctions cannot be so simply expressed. Because of the presence of the $\hat{\alpha}.\hat{p}$ term, the $\hat{s}$ and $\hat{l}$ operators do not commute with $\hat{H}$. The components of the $\hat{j} = \hat{l} + \hat{s}$ operator, however, do commute with $\hat{H}$, and so the eigenfunctions of the Hamiltonian may be eigenfunctions of $\hat{j}^2$ and $\hat{j}_z \cdot \hat{l}^2$ and $\hat{s}^2$ also commute with the Hamiltonian, and so we may still associate a particular *l* and *s* value with each solution, ψ.

It is useful, at this point, to express the four-component wave function as a pair of two component spinors, Φ and Θ :

$$\psi = \begin{pmatrix} \psi_1 \\ \psi_2 \\ \psi_3 \\ \psi_4 \end{pmatrix} = \begin{pmatrix} \Phi \\ \Theta \end{pmatrix}$$

Such a separation is dictated by the nature of the 4 × 4 Dirac spin matrices, and makes it possible to work with the familiar two-spinors of non-relativistic theory. Capitalizing on the commutation of $\hat{H}$ with $\hat{j}$, each spinor may be expressed as the product of a two-component eigenfunction of $\hat{j}$ which is dependent upon the spin and spatial angular coordinates and a radial function:

$$\Phi = g(r)X_I^{j,m_j} \text{ , } \Theta = f(r)X_I^{j,m_j}$$

The eigenfunctions of $\hat{j}$, X_I^{j,m_j}, may be expressed as sums of direct products of the more familiar eigenfunctions of and and the appropriate Clebsh-Gordon coefficients:

$$X_I^{j=(I+1/2),m_j} = X_{j,m_j}^{(+)} = \sqrt{\frac{l+m_j+1/2}{2l+1}}Y_I^{m_j-1/2}\left(\phi_{1/2}^{1/2}\right)$$

$$\sqrt{\frac{l-m_j+1/2}{2l+1}}Y_I^{m_j+1/2}\left(\phi_{1/2}^{-1/2}\right)$$

$$X_I^{j=(I-1/2),m_j} = X_{j,m_j}^{(-)} = -\sqrt{\frac{l-m_j+1/2}{2l+1}}Y_I^{m_j-1/2}\left(\phi_{1/2}^{1/2}\right) +$$

$$\sqrt{\frac{l+m_j+1/2}{2l+1}}Y_I^{m_j+1/2}\left(\phi_{1/2}^{-1/2}\right)$$

where $Y_I^{m_j\pm 1/2}$ are the spherical harmonics and

$$\phi_{\frac{1}{2}}^{\frac{1}{2}} = \alpha = \begin{pmatrix} 1 \\ 0 \end{pmatrix} \quad \phi_{-\frac{1}{2}}^{\frac{1}{2}} = \beta = \begin{pmatrix} 0 \\ 1 \end{pmatrix}.$$

Next, it is useful to define the operator

$$\hat{K} = (\sigma.\hat{l} + 1)$$

which commutes with $\hat{j}$. The eigenvalue equation associated with $\hat{K}$ is given by

$$\hat{K}X_{j,m_j}^{(\pm)} = \pm kX_{j,m_j}^{(\pm)}$$

where

$$k = \begin{cases} j + 1/2 & \text{if } j = l - 1/2 \\ -(j + 1/2) & \text{if } j = l + 1/2 \end{cases}$$

Because of the spherical symmetry of the nuclear potential, the eigensolutions may be further separated on the basis of their response to spatial coordinate inversion. The parity operator commutes with the full Hamiltonian, and so the final eigenfunctions, Ψ_{j,m_j}, must also obey the eigenvalue equation

$$\beta\Psi(-r) = \pm\Psi(r).$$

This necessarily dictates one of the following forms for the four-component eigensolutions:

$$\Psi_{j,m_j}^{(+)} = \begin{pmatrix} g(r)X^{(+)} \\ f(r)X^{(-)} \end{pmatrix} \quad \text{or}$$

$$\Psi_{j,m_j}^{(-)} = \begin{pmatrix} -g(r)X^{(-)} \\ f(r)X^{(-)} \end{pmatrix}$$

It is possible, through the use of the identity

$$\sigma.\hat{p} = (\sigma.\hat{r})\left(\hat{r}.\hat{p} + \frac{\sigma.\hat{l}}{r}\right)$$

and the property

$$(\sigma.\hat{r})\psi_{jm}^{(\pm)} = -\psi_{jm}^{\mp}$$

to completely separate out the angular functions from the eigenvalue equation, and thereby achieve two coupled, complex differential equations for the radial functions $f(r)$ and $g(r)$. The resultant equations are greatly simplified by the substitutions

$$r.f(r) = F(r) \text{ and } ir.g(r) = G(r)$$

to yield

$$(mc^2 - E - Z/r)F(r) - c\frac{dG(r)}{dr} - kcG(r)/r = 0$$

$$(-mc^2 - E - Z/r)G(r) + c\frac{dG(r)}{dr} - kcF(r)/r = 0$$

The solutions to these differential equations may be determined by first solving the asymptotic equations in the limit as r → ∞. These solutions, which are of the form

$$F(r) = \sqrt{1 + E/mc^2}\ \exp\left(\frac{-mc\sqrt{1-(E/mc^2)^2}}{\hbar}r\right),$$

$$G(r) = \sqrt{1 - E/mc^2}\ \exp\left(\frac{-mc\sqrt{1-(E/mc^2)^2}}{\hbar}r\right),$$

are then multiplied by an undetermined power series for both components and the bound state solutions are sought. The final form of each wave function is uniquely determined by the recursive relations of the power series and the boundary conditions, but may not be expressed in a simple general form. The resulting electronic energies are given by

$$E_n = mc^2\left[1 + (Z\alpha)^2 / \left(n - |k| + \sqrt{k^2 + (Z\alpha)^2}\right)^2\right]^{-1/2}$$

where α is the fine structure constant, and is given by $\frac{1}{c}$ in atomic units.

The four-component nature of the Dirac eigenfucntions gives rise to many interesting differences when compared to the non relativistic solutions. In order to better understand these differences, it is useful to inspect some simple approximations of the lowest energy hydrogenic solutions, for example the $n = 2$, $j = \{1/2, 3/2\}$ solutions.

$(n = 2, j = 1/2, k = 1, m_j = 1/2)$

$$\frac{1}{2\sqrt{2}}\begin{pmatrix} e^{-\rho/2}(1-\rho/2) \\ 0 \\ \frac{i}{2}Z\alpha e^{-\rho/2}(1-\rho/4)\cos\theta \\ \frac{i}{2}Z\alpha e^{-\rho/2}(1-\rho/4)\sin\theta e^{i\phi} \end{pmatrix}$$

$(n = 2, j = 1, k = 1, m_j = -1/2)$

$$\frac{1}{2\sqrt{2}}\begin{pmatrix} 0 \\ e^{-\rho/2}(1-\rho/2) \\ \frac{i}{2}Z\alpha e^{-\rho/2}(1-\rho/4)\cos\theta e^{i\phi} \\ -\frac{i}{2}Z\alpha e^{-\rho/2}(1-\rho/4)\cos\theta \end{pmatrix}$$

$(n = 2, j = 1/2, k = -1, m_j = 1/2)$

$$\frac{1}{4\sqrt{6}}\begin{pmatrix} e^{-\rho/2}\rho\cos\theta \\ e^{-\rho/2}\rho\sin\theta e^{i\phi} \\ -\frac{3i}{2}Z\alpha e^{-\rho/2}(1-\rho/6) \\ 0 \end{pmatrix}$$

$(n = 2, j = 1/2, k = -1, m_j = -1/2$

$$\frac{1}{4\sqrt{6}}\begin{pmatrix} e^{-\rho/2}\rho\cos\theta \\ e^{-\rho/2}\rho\sin\theta e^{i\phi} \\ 0 \\ -\frac{3i}{2}Z\alpha e^{-\rho/2}(1-\rho/6) \end{pmatrix}$$

$(n = 2, j = 3/2, k = 2, m_j = 3/2$

$$\frac{1}{8}\begin{pmatrix} e^{-\rho/2}\sin\theta e^{i\phi} \\ 0 \\ \frac{i}{4}Z\alpha e^{-\rho/2}(1-\rho/4)\sin\theta\cos\theta e^{i\phi} \\ \frac{i}{4}Z\alpha e^{-\rho/2}(1-\rho/4)\sin^2\theta e^{2i\phi} \end{pmatrix}$$

$(n = 2, j = 3/2, k = 2, m_j = -3/2)$

$$\frac{1}{8}\begin{pmatrix} 0 \\ e^{-\rho/2}\sin\theta e^{i\phi} \\ \frac{i}{4}Z\alpha e^{-\rho/2}(1-\rho/4)\sin^2\theta e^{2i\phi} \\ -\frac{i}{4}Z\alpha e^{-\rho/2}(1-\rho/4)\sin\theta\cos\theta e^{i\phi} \end{pmatrix}$$

$(n = 2, j = 3/2, k = 2, m_j = 1/2)$

$$\frac{1}{4\sqrt{3}}\begin{pmatrix} e^{-\rho/2}\sin\theta e^{i\phi} \\ -\frac{1}{2}e^{-\rho/2}\sin\theta e^{i\phi} \\ -\frac{i}{4}Z\alpha e^{-\rho/2}\rho(1-3\cos^2\theta) \\ \frac{3i}{8}Z\alpha e^{-\rho/2}\rho\sin\theta\cos\theta e^{i\phi} \end{pmatrix}$$

$(n = 2, j = 3/2, k = 2, m_j = -1/2)$

$$\frac{1}{4\sqrt{3}}\begin{pmatrix} \frac{1}{2}e^{-\rho/2}\sin\theta e^{i\phi} \\ e^{-\rho/2}\sin\theta e^{i\phi} \\ \frac{3i}{8}Z\alpha e^{-\rho/2}\rho\sin\theta\cos\theta e^{i\phi} \\ \frac{i}{8}Z\alpha e^{-\rho/2}\rho(1-3\cos^2\theta) \end{pmatrix}$$

These solutions are correct to order αZ, as illustrated in Moss's discussion of the hydrogenic solutions to Dirac's equation. Here, analogous to the non-relativistic case, for each unique combination of the principal quantum number n and a specific angular momentum quantum number j, the solutions must be further classified according to the projection of their total angular momentum on the z-axis, m_j.

These solutions are, of course, degenerate in the absence of an external field. The contributions from the first two components of the wave-function tend to be much larger than the contributions from the final two components for electron-like solutions. For this reason the upper two-component spinor is known as the large component of the wave-function, while the lower spinor is known as the small component. By orthogonality, the opposite is true for the positron-like solutions. Since the angular and spin portions of the wave-functions are normalized, the magnitude of the large and small contributions is determined entirely by the radial functions, $f(r)$ and $g(r)$. Given the asymptotic form of the radial equations and the orbital energy, the ratio of their contributions may be expressed as

$$\frac{g(r)}{f(r)} \simeq \frac{Z\alpha}{2n}$$

In general, the ratio will probably be larger than this approximate value in regions close to the nucleus, where the contribution of $g(r)$ is the greatest. The contribution of the small component, then, may be significant for sufficiently heavy nuclei. Though hydrogenic ions with very heavy nuclei may not be of great practical interest, this relationship will have implications for heavy many-electron atoms where the innermost electrons experience a large portion of the full nuclear charge and hence are closely related to the analogous hydrogenic systems.

One interesting consequence of the four component wave-function may be observed in the associated probability density. Scalar wave-functions, ψ, which possess radial and angular nodes will have the same radial and angular nodes in their probability density, $\psi^*\psi$ The four-component wave-function, however, gives rise to a probability density which is the sum of the probability densities associated with the large and small wave-functions, ϕ and χ, respectively.

The radial component of ϕ and χ posses different numbers of nodes and, in general, none of these nodes will coincide. Similarly, the angular contributions of ϕ differ from those of χ by a single unit of angular momentum. Therefore, although $\phi^*\phi$ and $\chi^*\chi$ may possess either radial of angular nodes individually, their sum, their sum will be node-less. This is illustrated by Figure, where the large and small component radial functions for the $n = 2, j = 1/2, k = -1$ orbitals, $g(r)$ and $f(r)$, have been plotted, along with the resultant radial probability density, $\rho(r) = g(r)^2 + f(r)^2$.

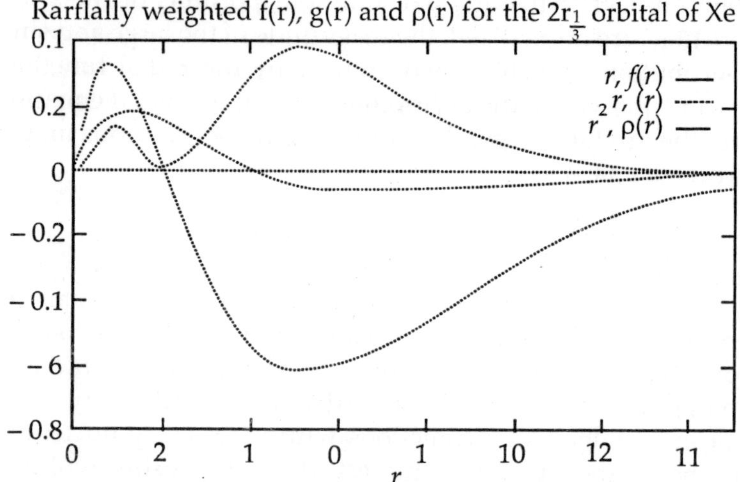

The energy eigenvalues of the hydrogenic solutions to the Schrödinger equation are only dependent upon n, the principal quantum number, while the Dirac hydrogenic eigenvalues are dependent on both n and j. The spectral dependence on j is upheld by experimental observation, however the degeneracy of eigenvalues for solutions with the same j values but differing l values is not observed in nature. The breaking of this degeneracy is known as the Lamb shift and its origin has been attributed to the difference in the interaction of the different j eigenfunctions with the vacuum fluctuations predicted by quantum electrodynamics.

The magnitude of the Lamb shift is much smaller than the splitting introduced by spin orbit coupling and is largest for the $n = 2, j = 1/2$ shells of the hydrogenic atom. In this case, the splitting introduced by the Lamb shift is approximately 10% of the magnitude of the energy separation of the $p_{1/2}$ and $p_{3/2}$ eigenfunctions. In order to understand the source of the Lamb shift, it is necessary to appeal to quantum electrodynamics, where higher order couplings of the electromagnetic interactions of the electron and nuclei are taken into account.

Electron Spin Angular Momentum and Magnetic Moment

One of the triumphs of the Dirac equation was its explicit connection to electron spin. The Dirac equations does not predict electron spin as a relativistic property, observable only under relativistic conditions. Reduction of the Dirac equation for an electron in a magnetic field to its non-relativistic limit yields the Schrödinger equation with a correction term which takes account of the interaction of the electron's intrinsic magnetic moment with the magnetic field. Using the definition of the Dirac Hamiltonian, $\hat{H}_D$, given in, we may arrive at the equation

$$(\hat{H}_D + e\phi)^2 / c^2 = (\beta mc + \alpha.\pi)^2$$

Due to the anti-commutation properties of the components of α, we may expand the right hand side to reveal that

$$(\hat{H}_D + e\phi)^2 c^{-2} = m^2c^2 + (\alpha.\pi)(\alpha.\pi)$$

If we define α_i our matrices as products of a matrix ρ and σ_{i4} the matrices where

$$\rho = \begin{pmatrix} 0 & I_{2x2} \\ I_{2x2} & 0 \end{pmatrix} \text{ and } \sigma_{i4} = \begin{pmatrix} \sigma_i & 0 \\ 0 & \sigma_i \end{pmatrix}$$

so that $\alpha_i = \rho \otimes \sigma_{i4}$, then we may express $(\alpha.\pi)$ $(\alpha.\pi)$ as

$$(\alpha.\pi)(\alpha.\pi) = (\alpha.\pi)(\alpha.\pi)$$

where we have used the identity ρ^2 = I. This may be rearranged to the form

$$(\alpha.\pi)(\alpha.\pi) = \pi.\pi + i\sigma(\pi \times \pi)$$

via sundry vector identities. The cross-product of π with itself does not drop out. This is because the components, π_i, of the vector π are not simple scalars, but rather the sum of a differential operator, p^i, and a scalar, A^i, so the cross terms of the cross products do not vanish, making the cross-product of π with itself

$$(\pi \times \pi) = -i\hbar\nabla \times A = -i\hbar B$$

Returning to our original equation, we may make this substitution to arrive at

$$(\widehat{H}_D + e\phi)^2 c^{-2} = m^2c^2 + \pi^2 + \hbar\sigma_4 B$$

In the non-relativistic limit this gives us a Hamiltonian of the form

$$\widehat{H}_D = -e\phi + \frac{\pi^2}{2m} + \frac{\hbar}{2m}\sigma B$$

The final term indicates that a term which corresponds to a body with a magnetic moment of $-\frac{\hbar}{2m}\sigma$ in a magnetic field B must be added to give the proper energy. This prediction of electron spin as a property which should exist in the non-relativistic limit provides strong evidence that Dirac's equation provides a more complete physical picture than the wave equation of Schrödinger, where electron spin must be treated in an *ad hoc* manner.

DIRAC-HARTREE-FOCK METHOD

The use of the Dirac Hamiltonian as a relativistic substitute for the one electron terms of the traditional, non-relativistic many-electron Hamiltonian yields the Dirac-Coulomb Hamiltonian: one of the most widely used special-relativistic, many-electron Hamiltonians. This Hamiltonian may be utilized in conjunction with a Hartree-Fock-like wave-function in what is known as the Dirac-Hartree-Fock (DHF) method.

The DHF method has a special status in quantum chemistry as it often is utilized to benchmark relativistic effects in the absence of electron correlation. Such benchmarks can provide a gauge of the accuracy of more approximate methods which attempt to include relativistic effects as a perturbation of the non-relativistic Hartree-Fock case as well as methods which employ transformed, simplified versions of the Dirac-Coulomb Hamiltonian.

A MANY ELECTRON HAMILTONIAN

The Dirac equation has been shown to successfully treat the interactions of electrons with nuclei in a special-relativistic manner, and so represents a good starting point for a special-relativistic many-electron Hamiltonian. What is needed next is a description of the electron-electron interaction, and an associated quantum mechanical operator, $\hat{g}_{ij}$. If the electron-electron interaction is not altered significantly by the introduction of special relativity, then the standard, non-relativistic coulomb operator, $\hat{g}_{ij} = \frac{1}{r_{ij}}$, might not be a bad guess. With the coulomb interaction, the Hamiltonian for an n electron system would be given by

$$\hat{H} = \sum_{i=1}^{n} \hat{h}_D^i + \sum_{i>j}^{n} \frac{1}{r_{ij}}$$

$$= \sum_{i=1}^{n} (-e\phi(r_i) + c\hat{\alpha}^i.\hat{p}^i + mc^2) + \sum_{i>j}^{n} \frac{1}{r_{ij}}$$

This is known as the Dirac-Coulomb Hamiltonian, $\hat{H}_{DC}$. The associated wave equation, $\hat{H}_{DC}\Psi = E\Psi$, is not Lorentz invariant, and so $\hat{H}_{DC}$ does not represent a proper special relativistic Hamiltonian. In order to obtain a two-electron interaction term which is consistent with special relativity, it is necessary to turn to quantum electrodynamics (QED). In order to cast the QED electron-electron interaction term into a reasonable form, it is necessary to expand it in perturbative series in orders of the fine structure constant, α. Retaining only the terms which contribute up to order α^2, gives the Coulomb operator plus the Breit interaction term

$$\hat{g}_{ij}^B = \frac{-1}{r_{ij}} \left(\alpha_i.\alpha_j + \frac{(\alpha_i.r_{ij})(\alpha_j.r_{ij})}{r_{ij}^2} \right)$$

In practice, however, the full Breit magnetic interaction term is often cumbersome to implement, and so an approximation of the Breit operator known as the Gaunt operator, may be used

$$\hat{g}_{ij}^{G} = \frac{-\alpha_i . \alpha_j}{r_{ij}}.$$

The Gaunt operator is not gauge invariant, but it avoids having to solve integrals over operators more complicated than $\frac{1}{r_{ij}}$, and includes the largest contributions of the Breit interaction.

More rigorous forms of the molecular Hamiltonian have been suggested, but they are, at best, only approximately Lorentz covariant to some order in α. Because the two-electron magnetic interaction terms are typically small, the contributions beyond second order in α are typically chemically unimportant. Because of this the more rigorous Hamiltonians, which typically involve more complicated operators than even the Breit interaction, have received far less attention from quantum chemical investigators, and so the Dirac-Coulomb, Dirac-Coulomb-Gaunt, and Dirac-Coulomb-Breit Hamiltonians are the most widely used four-component special relativistic molecular Hamiltonians.

ATOMIC SOLUTIONS

Atomic eigenfunctions of the D-C Hamiltonian may not be achieved analytically as were the hydrogenic solutions of the single-particle Dirac equation. Instead, it is necessary, as it was in the non-relativistic theory, to appeal to approximate models such as the Dirac-Hartree-Fock (DHF) method. In a manner analogous to Hartree-Fock theory, DHF begins with the assumption that an n-electron atomic wavefunction, Ψ, can be represented by an antisymmetrized product of n single particle functions, $\{\phi_i\}$. In contrast to the HF method and in accordance with the 4-component nature of the single particle operators, $\hat{h}_D^i$, the single particle functions ϕ_i are 4-component spinors. If we make the approximation that each electron experiences a central potential only, then the single particle functions may be separated into radial, angular, and spin parts analogous to the hydrogenic eigenfunctions.

Anticipating the existence of electronic and positronic solutions, we may propose two separate forms for ϕ_i, analogous to the approximations of the hydrogenic solutions presented above:

$$\phi_i^{j,m_j}(+) = \frac{1}{r}\begin{pmatrix} p_i(r)\chi_i^{(+)} \\ iq_i(r)\chi_i^{(-)} \end{pmatrix} \text{ and}$$

$$\phi_i^{j,m_j}(-) = \frac{1}{r}\begin{pmatrix} -p_i(r)\chi_i^{(-)} \\ iq_i(r)\chi_i^{(+)} \end{pmatrix}$$

where $\chi_i^{(\pm)}$ are two component spinors dependent upon the spin and angular coordinates and the quantum numbers l, j and m_j, and are given by the $\hat{j}$ eigenfunctions ψ_I^{j,m_j} presented for the hydrogenic case in Section. The form of the radial functions has, again, been chosen such that the resultant wave equations and electronic energy expression are simplified. The total wavefunction formed by these one particle functions may be expressed as

$$\Psi = A\left(\prod_{i=1}^{n}\phi_i(r_i)\right)$$

$$= \frac{1}{\sqrt{n!}}\begin{vmatrix} \phi_1(r_1) & \phi_2(r_1) & \cdots & \phi_n(r_1) \\ \phi_1(r_2) & \phi_2(r_2) & \cdots & \phi_n(r_2) \\ \vdots & \vdots & & \vdots \\ \phi_1(r_n) & \phi_2(r_n) & \cdots & \phi_n(r_n) \end{vmatrix}$$

where A is the antisymetrizer operator. If this form of the atomic wavefunction is assumed, and the Dirac-Coulomb Hamiltonian is utilized, then the electronic energy is given by

$$E = \left\langle \Psi \,|\, \hat{H}_{DC} \,|\, \Psi \right\rangle$$

$$= \sum_{i}^{n} \langle \phi_i \mid \hat{h}_d \mid \phi_i \rangle + \frac{1}{2} \sum_{i,j}^{n} \left(\left\langle \phi_i \phi_j \left| \frac{1}{r_{12}} \right| \phi_i \phi_j \right\rangle - \left\langle \phi_i \phi_j \left| \frac{1}{r_{12}} \right| \phi_i \phi_j \right\rangle \right)$$

In order to obtain solutions of the form, it is necessary to derive a Fock-type operator, which, like its non-relativistic counterpart, has employed the central field approximation in order to define an averaged radial potential operator, $V(r)$ to replace the exchange and Coulomb operators. This potential will still be dependent upon the form of the single particle functions, $\{\phi_i\}$, and must be solved for iteratively, in a self-consistent manner. Such a potential is presented by Grant in his extensive discussion of relativistic atomic structure.

The resultant Fock operators may be derived by requiring that the form of the occupied single particle spinors are variationally optimized with respect to the electronic energy. In the variational construction of the Dirac-Fock operator, it is important to be sure that the occupied ϕ_i represent purely electronic solutions.

If positronic states are occupied, instead, the single particle spinors will collapse into the negative energy continuum. In the case that the atomic DHF wavefunction is to be represented on a numerical grid, some relatively straightforward boundary conditions may be applied in the occupation selection procedure to be sure that the final wavefunction is not contaminated by positronic solutions.

This method of solution has been implemented in a variety of widely available programmes, most notably the programme due to Desclaux.

MOLECULAR SOLUTIONS: FINITE BASIS SET EXPANSION

The application of the Hartree-Fock method of obtaining approximate molecular wavefunctions for this choice of Hamiltonian proceeds in a manner similar to that used to obtain atomic HF wavefunctions. Once again, the total electronic wavefunction is represented by an antisymmetrized product of single particle wavefunctions, $\{\phi_i\}$

$$\Psi_e = A\frac{1}{\sqrt{n!}}\left|\prod_i \phi_i\right|$$

For molecular systems, the separation of radial and angular portions of the wavefunction is no longer possible, and so a numerical representation of Ψ would require that each ϕ_i be represented on an extensive 3-dimensional grid of points. The evaluation of two-center integrals over this grid becomes impractically expensive for systems of even modest size. Instead, a different sort of finite basis set must be chosen to represent the single particle wavefunctions.

Traditionally, the view of molecular electronic structure has been built upon the better-understood framework of atomic electronic structure. Therefore, it is not surprising that the most popular basis sets employed have been closely related to the hydrogenic solutions which form the framework for the understanding of many-electron atoms. For solving the non-relativistic wave equation, it would appear that perhaps a good basis set would be comprised of functions $\{\phi_\mu\}$ which are centered on the constituent atoms of the molecular system of interest with a functional form given by

$$\phi_\mu^a = e^{-ar_a} Y_l^m(\theta_a, \phi_a)$$

where the index **a** denotes the atomic center to which the spherical coordinates refer, and the exponential scale factor, α, may be optimized for the atomic system of interest. This choice of basis functions, known as Slater type functions (STF), makes solution of the multi-center electron repulsion integrals which arise in the HF equations extremely difficult to solve accurately. For this reason, the radial portion of the basis functions is more commonly represented by Gaussian type functions (GTF) to yield the basis functions

$$\phi_\mu^a = e^{-ar_a^2} Y_l^m(\theta_a, \phi_a)$$

where, again, the α's may be optimized for the atomic system of interest. This choice of basis set does have significant drawbacks, however, as is evidenced in the ability of this sort of basis set to represent the ground state of the hydrogen atom. Replacing the exact wavefunction of the 1s electron of ground state H with a single normalized GTF with a variationally optimized exponent, α, yields an electronic energy which is in error by 15%. The use of a linear combination of n normalized GTF's

$$\psi_{1s}^{appx.} = \sum_{\mu}^{n} C_\mu N_\mu e^{-a_\mu r^2} y_0^0$$

with variationally optimized C_μ and α_μ yields smaller errors for higher values of n: 2.8% for $n = 2$, 0.15% for $n = 4$, 0.016% for $n = 8$. The reason for the inadequacy of the GTF's is directly associated with their asymptotic behaviour, both at the nucleus, and as $r \to \infty$.

Although some attempts have been made to implement techniques which employ more compact basis sets constructed solely of STF's, the extra work associated with optimizing the various parameters associated with the GTF basis has proven to be far more manageable than the difficulties which arise with solving electron repulsion integrals involving STF's.

The form of the atomic solutions of the DHF equations points toward the selection of a four-component basis which is analogous to equation. Because two-electron integrals analogous to the Coulomb and exchange integrals of the HF equations also arise in the DHF equations, it is necessary to instead utilize a basis set analogous to equation.

Because the large and small components of the DHF wavefunction exhibit distinct roles in the description of molecular wavefunctions, it is useful to describe each with a separate basis set. Furthermore, it will prove to be useful to construct the basis functions such that the two component large and small basis spinors are all eigenfunctions of $\hat{\sigma}_{4z}$, so that the large and small basis sets are given by

$$\chi_\mu^{4L} = \begin{pmatrix} \chi_\mu^L \\ 0_2 \end{pmatrix} \text{ and } \chi_\mu^{4S} = \begin{pmatrix} 0_2 \\ \chi_\mu^S \end{pmatrix}$$

where the two component spinors, $\{\chi_\mu^X\}$ are the product of a function of the spatial coordinates, ϕ_μ^X, and one of the eigenvectors of $\hat{\sigma}_z$:

$$\alpha = \begin{pmatrix}1\\0\end{pmatrix} \text{ or } \beta = \begin{pmatrix}0\\1\end{pmatrix}.$$

In a molecular calculation, linear combinations of Cartesian Gaussian type functions (CGTF) centered on the a particular nucleus are typically employed as spatial basis functions in lieu of the products of radial functions and spherical harmonics which the atomic solutions would appear to suggest as a logical basis. A general CGTF is given by

$$f(x,y,z) = N(i,j,k,\alpha)x^i y^j z^k e^{-\alpha r^2}.$$

This choice of basis greatly simplifies the analytic evaluation of integrals involving these functions via the Gaussian product theorem and various recursive relationships which exist among CGTFs. Functions of this type with $i+j+k = q$ may be classified as having angular momentum q, since linear combinations of each group of 3^q CGTFs with the same q value give rise to $2q + 1$ functions which are directly related to the solid spherical harmonics of angular momentum q: $r^q Y_{qm}(\theta,\phi)$. The transformation to this set of "pure angular momentum" basis functions does not affect the evaluation of integrals involving these functions, though it can affect the size of the basis set, since, for $q > 1$, the number of CGTFs is greater than the number of pure angular momentum functions. The molecular single particle 4-spinors, ϕ_i, may be approximated by a linear combination of the large and small component basis functions centered on the molecule's constituent atoms:

$$\phi_i = \sum_\mu^{N^L} C_{\mu\alpha}^{iL}\phi_{\mu\alpha}^{4L} + \sum_\mu^{N^L} C_{\mu\beta}^{iL}\phi_{\mu\beta}^{4L} + \sum_\mu^{N^S} C_{\mu\alpha}^{iS} i\phi_{\mu\alpha}^{4S} + \sum_\mu^{N^S} C_{\mu\beta}^{iS} i\phi_{\mu\beta}^{4S}$$

where N^L and N^S are the number of large and small component basis functions, respectively. Adopting a shorthand notation, which will allow simplification of expressions involving the evaluation of operators in the basis, the 4-spinor ϕ_j becomes j while the two- and four-spinor basis functions $\phi_{\mu s}^{X}$ and $\phi_{\mu s}^{4X}$ become μ_s^X and μ_s^{4X}, respectively, where X is either L or S, and s is either α or β. The value of the DHF electronic energy, first presented in equation, may be expressed in this new shorthand as

$$E_{DHF} = \sum_{i}^{n_c} \langle i | \hat{h}_d | i \rangle + \frac{1}{2} \sum_{ij}^{n_c} \langle ij || ij \rangle.$$

The next task is to derive a set of Fock-like equations from this Energy expression. If we express ϕ_i as a sum of the most general sort of 4-spinor basis function, ϕ_{μ}^{4X}, it becomes possible to express the DHF energy in terms of these orbitals and their expansion coefficients, C_{μ}^{4X}, so that

$$E_{DHF} = \sum_{i}^{n_c} \sum_{X,Y}^{L,S} \sum_{s,p}^{\alpha,\beta} \sum_{\mu}^{n^X} \sum_{y}^{n^Y} C_{\mu s}^{riX*} C_{vp}^{riY*} \langle \mu_s^{4X} | \hat{h}_D | V_p^{4X} \rangle$$

$$+ \frac{1}{2} \sum_{i}^{n_c} \sum_{W,X,Y,Z}^{L,S} \sum_{s,p,q,r}^{\alpha,\beta} \sum_{\mu}^{N^W} \sum_{\rho}^{N^X} \sum_{\sigma}^{N^Z}$$

$$C_{\mu s}^{ri} * C_{vp}^{j} * C_{\rho q}^{i} C_{\sigma r}^{j} \langle \mu_s^{4W} v_p^{4X} || \rho_q^{4Y} \sigma_r^{4Z} \rangle.$$

The DHF energy expression may be more compactly expressed through the use of the intermediates

$$P_{\mu v}^{XY,sp} = \sum_{i}^{n_c} C_{\mu s}^{iX} * C_{vp}^{iY},$$

$$G_{\mu v \rho \sigma}^{WXYZ,spqr} = \sum_{ij}^{n_c} \langle \mu_s^{4W} v_p^{4X} | \rho_q^{4Y} \sigma_r^{4Z} \rangle - \langle \mu_s^{4W} v_p^{4X} | \sigma_r^{4Z} \rho_q^{4Y} \rangle$$

to yield

$$E_{DHF} = H_{\mu v}^{XY,sp} P_{\mu v}^{XY,sp} + \frac{1}{2} P_{\mu\rho}^{WY,sq} G_{\mu v\rho\sigma}^{WXYZ,spqr} P_{v\sigma}^{XZ,pr}$$

where the repeated indices are assumed to be summed. This energy expression, along with the imposition of the orthonormality condition

$$\langle i \mid j \rangle = \sum_{\mu v}^{N^X, N^Y} \sum_{s,p}^{\alpha\beta} \sum_{X,Y}^{L,S} \langle \mu_s^{4X} \mid v_p^{4Y} \rangle = \delta_{ij}$$

Allows for the construction of the Lagrangian

$$L[C_\mu^{iX}] = E_{DHF} - \sum_{ij}^{n_c} \epsilon_{ij} \left[\langle i \mid j \rangle - \delta_{ij} \right].$$

This Lagrangian may subsequently be subjected to the condition that the first derivative of L with respect to the orbital expansion coefficients, C_μ^{iX}, be equal to zero. This method of variationally solving for the optimal Dirac-Hartree-Fock wavefunction is completely analogous to the method put forth by Roothaan for the non-relativistic case. The imposition of the stationary condition gives rise to a set of linear equations which may be expressed as

$$F_{DC}^{YX,ps} C_v^{iX,s} = \epsilon_i \, S_{\mu v}^{XY,ps} C_v^{iY,s}$$

where

$$S_{\mu v}^{XY,sp} = \langle \mu_s^{4X} \mid v_p^{4Y} \rangle,$$

and the matrix elements of the Dirac-Fock-Roothaan operator $\hat{F}^{DC}$ in the GTO basis are given by

$$F_{DC\mu v}^{XY,sp} = H_{\mu v}^{XY,sp} + G_{\mu\rho v\sigma}^{XWYZ,srpq} P_{\rho\sigma}^{WZ,rq}.$$

The DC subscript on the Dirac-Fock operator acknowledges that this operator is based upon the choice of the Dirac-Coulomb Hamiltonian. For these equations to be useful, however, they must be cast in terms of integrals involving only the scalar contracted CGTF basis functions, $\{\phi_\mu^X\}$.

The matrix elements of the one-particle Dirac Hamiltonian in the basis are given by

$$\left\langle \mu_s^{4X} \mid \hat{h}_D \mid v_p^{4Y} \right\rangle = H_{D\mu v}^{XY,sp}$$

$$= V_{\mu v}^{XY,sp} + \Pi_{\mu v}^{XY,sp}$$

A significant number of these terms integrate to zero once the explicit form of the basis functions is inserted, and the matrix element are evaluated. For the V integrals, only those matrix elements involving two large component basis functions with the same spin symmetry or two small component basis functions with the same spin symmetry are non-zero, so that

$$V_{\mu v}^{XY,sp} = \delta_{XY}\delta_{xp} \left\langle \mu^X \mid \hat{V}_{nuc} \mid v^Y \right\rangle$$

where

$$\hat{V}_{nuc} = \sum_a^{N_{nuc}} \frac{Z_a}{r_a}$$

For the Π integrals, on the other hand, only the matrix elements between a large and small basis function will have a non-zero value. Integration over the spin variables leaves elements which are non-zero for any combination of spins on the two basis functions:

$$\Pi_{\mu v}^{XY,sp} = (1-\delta_{XY})\Big[\left\langle \mu^X \mid \lambda_p \nabla_z \mid v^Y \right\rangle \delta_{sp}$$

$$+ \left\langle \mu^X \mid \nabla_x - i\lambda_p \nabla_y^{|} v^Y \right\rangle (1-\delta_{sp})\Big]$$

The overlap integrals may be reduced in a similar fashion:

$$S_{\mu v}^{XY,sp} = \delta_{XY}\delta_{xp} \left\langle \mu^X \mid v^Y \right\rangle$$

Finally, reduction of a general four-component two-electron integral to integrals over the scalar spatial basis functions yields

$$\left\langle \mu_s^{4W} v_p^{4X} \left| \frac{1}{r_{12}} \right| \rho_q^{4W} \sigma_r^{4X} \right\rangle = \delta_{WY}\delta_{XZ}\delta_{sq}\delta_{pr} \left\langle \mu^W v^X \left| \frac{1}{r_{12}} \right| \rho^W \sigma^X \right\rangle.$$

This reduction of the pertinent integrals makes it possible to take advantage of the lore which has accumulated regarding the efficient evaluation of non-relativistic electronic integrals. Since the basic quantities which must be evaluated are the same as for the non-relativistic case, highly optimized non-relativistic integral codes may be adapted to evaluate relativistic integrals.

SELECTION OF THE BASIS

Because of the presence of the continuum of positronic states in the DHF model, it is important to be certain that only electronic states spinors are occupied in the construction of the relativistic Fock matrix. One way of unambiguously imposing this restriction is via the method of kinetic balance. In regions where the nuclear potential is small relative to the rest mass energy, the single particle Dirac equation predicts that, for positive energy solutions, the small component is approximately related to the large component by means of the relation

$$\phi_S = \frac{1}{2mc}\sigma.\hat{p}\phi_L.$$

This proportionality is a direct result of the form of the kinetic energy operator, $\alpha.\pi$. This condition should be obeyed by electronic solutions in regions far from the nucleus. If the basis set does not reflect this relationship between the two components, positronic states will contaminate the electronic solutions, and a limited variational collapse will ensue. This problem may be remedied by first selecting a large component basis, and then constructing a small component basis by applying the $\sigma.\hat{p}$ operator to each large component.

The resultant basis set, which is said to be kineticly balanced, will be able to fulfill the low field boundary conditions without requiring mixture of the positive and negative energy solutions. The action of the kinetic energy operator on the atomic basis functions has some interesting consequences. $\alpha.\pi$ acting on a large component *s*-type Gaussian function gives rise to a *p*-type small component Gaussian function. The action of the kinetic energy operator on a large component *p*-type Gaussian gives rise to a small component *s* function and a small component *d* function. In general, a large component atomic basis basis function of angular momentum *l* gives rise to small component atomic orbitals of angular momentum *l*-1 and l + 1. As a result, the small component basis can be up to almost twice as large as the large component basis. For this reason, a relativistic basis set can involve almost three times as many basis functions as a non-relativistic basis of similar quality.

Once a kineticly balanced basis set has been constructed and the scalar integrals have been evaluated, the DHF equations may be implemented in a straightforward manner. In order to improve the efficiency of these calculations, however, several aspects of the symmetry of the electronic wavefunction may be exploited. In non-relativistic quantum chemistry, the use of linear combinations of basis functions which transform as irreducible representations of the point group of the nuclear framework's symmetry can dramatically reduce the number of integrals which need to be solved, and hence the size of the Fock matrix, since entire blocks of the Fock matrix become zero by symmetry. Spin-restriction is another common method of simplifying the evaluation of the non-relativistic HF wavefunction.

When spin-restriction is imposed, the same set of single particle functions, $\{\phi_i\}$, is used to describe the spatial portion of electrons with α and β spin. This allows for the reduction of the size of the Fock matrix, and results in a wavefunction which is rigorously an eigenfunction of $\hat{S}^2$. Spatial symmetry constraints and spin-restriction may be performed simultaneously for non-relativistic methods.

The symmetry properties of the relativistic single particle spinors, on the other hand, must be considered within the framework of double-groups. Double groups take into account all of the spatial symmetry features which the spatial symmetry point groups consider plus rotations by 2π. Rotations by 2π introduce a sign flip for fermionic wavefunctions, and so this adds an additional criterion which may be used to characterize the symmetry of the wavefunction. This extra characteristic doubles the order of the group, hence the appellation "double-group symmetry".

In order to take advantage of this type of symmetry, symmetry adapted basis functions which transform as a particular irreducible representation of the double group of interest are constructed from linear combinations of the original basis functions. All of the scalar integrals are then constructed using this symmetry adapted basis, thereby lending the same symmetry constraints to the final DHF wavefunction.

Another consequence of the entanglement of the spin and spatial symmetry is that it is not possible to apply spin restriction in the same manner as it was applied for non-relativistic wavefunctions. Instead, the related Kramer's restriction may be imposed. Kramer's restriction relates two spinors through the time reversal operator, $\widehat{K}_4$. The time reversal operator is given by

$$\widehat{K}_4 = -i\begin{pmatrix} \sigma_y & 0 \\ 0 & \sigma_y \end{pmatrix}\widehat{K}$$

where K is the scalar complex conjugation operator, ϕ_i and ϕ_i relates a pair of spinors, and, via the relations

$$\widehat{K}\phi_i = \overline{\phi_i} \text{ and } \widehat{K}\overline{\phi_i} = -\phi_i$$

The degeneracy of the Kramer's pairs reduces the number of integrals and matrix elements which need to be evaluated to construct the DHF operator.

DHF-BASED CORRELATED METHODS

The DHF method represents a relatively rigorous technique for obtaining a single determinant wavefunction which accounts for the most important consequences of special relativity. For many interesting chemical compounds, however, a single-determinant wavefunction is insufficient, and the inclusion of many-body interaction terms, which are not taken into account in the Hartree-Fock approximation, are essential for a proper description of molecular electronic structure. Much of the effort in quantum chemistry has gone into the design and implementation of methods which are able to effectively and efficiently account of these many-body terms. Methods such as multi-configuration self-consistent-field (MCSCF) theory, many-body perturbation theory (MBPT), configuration interaction (CI) theory, and coupled-cluster (CC) theory, the four most popular so-called correlated methods, have been designed expressly to address the many-body problem.

Because the many-body problem does not disappear for systems which require relativistic treatment, the DHF model may not be able to provide highly accurate predictions. Taking a cue from non-relativistic quantum chemistry, MCSCF, MBPT, CI, and CC methods have been adapted for use with a DHF reference wavefunction. While the DHF-based correlated methods can be extremely computationally expensive, they represent the most rigorous methods currently available for accurate prediction of molecular electronic structure, and may serve both as computational bench marks and extremely reliable tools for the treatment of systems which exhibit significant dependence on both correlation and relativistic effects.

Solution of the Dirac-Fock equations for a molecular system in a finite basis set expansion gives rise to set of orthonormal single particle 4-spinor solutions, $\{\phi_i\}$, and a Dirac-Fock wavefunction, Ψ_{DF}, which is an antisymmetrized product of a subset of these single particle states, typically known as a Slater determinant.

The spinors which are used to construct the Dirac-Fock Ψ_{DF} are classified as occupied while the remainder are considered unoccupied. In the solution of the Dirac-Fock equations, the occupied states are the N Ψ_{DF} single particle electronic eigenstates of the Dirac-Fock operator with the lowest eigenvalues, where N Ψ_{DF} is the number of electrons. The positronic states, of course, always posses lower eigenvalues than any of the electronic states, but are excluded from the Ψ_{DF} to avoid variational collapse.

Using both the occupied and unoccupied single particle states, it is possible to construct a large number of determinantial wavefunctions to represent the total electronic wavefunction. In order for these determinants to be suitable descriptions of the electronic wavefunction for the system of interest, they must each be an antisymmetrized product of $N_\in$ distinct spinors. These determinants may be classified with respect to the number of spinors they have in common with the Dirac-Fock wavefunction.

If a particular determinant differs from the DF wavefunction by only one spinor, it is considered a singly-excited determinant, since it may be thought of as the DF wavefunction for a molecular system which has one electron excited relative to the system considered in the original DF optimization. Doubly-excited determinants differ from the DF wavefunction by two spinors, triply-excited determinants differ by three, and so on. Excited determinants may be classified by the specific orbital replacements which distinguish them from the DF or reference determinant. Singly excited derminants may be denoted by a label such as S_i^a, which specifies that the occupied spinor ϕ_i has been replaced by the unoccupied spinor ϕ_a, while doubly excited determinants would be labeled by D_{ij}^{ab}, etc. Correlated methods seek to utilize mixtures of these excited determinants to improve the quality of the approximate wavefunction, and the accuracy of the predicted electronic energy, by constructing a wavefunction which includes contributions from these determinants in some fashion.

Configuration Interaction

One of the most conceptually straightforward methods of obtaining a better wavefunctions is given by the configuration interaction (CI) method. The CI method employs a wavefunction which is constructed of a linear combination of the DHF wavefunction with excited determinants

$$\Psi^{CI} = C_0 \Psi^{DHF} + \sum_{i}^{acc} \sum_{a}^{vir} C_i^a s_i^a + \sum_{ij}^{acc} \sum_{ab}^{vir} C_{ij}^{ab} D_{ij}^{ab} + \ldots,$$

The expansion coefficients are then selected so that they variationally minimize the expectation value of the electronic energy with respect to the CI wavefunction

$$E_{CI} = \langle \Psi_{CI} | \hat{H} | \Psi_{CI} \rangle$$

where the wavefunction is restricted by the normalization condition

$$\begin{aligned} 1 &= \langle \Psi_{CI} | \Psi_{CI} \rangle \\ &= C_0^2 + \sum_{i}^{acc} \sum_{a}^{vir} C_i^a + \sum_{ij}^{acc} \sum_{ab}^{vir} C_{ij}^{ab^2} + \ldots \end{aligned}$$

In the most common implementation of this method, the wavefunction is constructed only from certain classes of excited determinants. The methods used to select the best set of excited determinants, including a great deal of recent work in the area of restricted active space (RAS) selection schemes, have been the subject of much discussion.

The wavefunction which has been constructed from all the possible excited determinants is known as the full CI (FCI) wavefunction, and represents the exact solution of the DC Hamiltonian within the space constructed of the electronic DHF spinors. An efficient version of a DHF based RAS CI programme has been implemented by Visscher and coworkers in the MOLFDIR programme package.

MULTI-CONFIGURATION SELF CONSISTENT FIELD METHOD

In the Dirac-Fock method, the form of the single particle functions constructing Ψ_{DF} was variationally optimized by selecting expansion coefficients C_μ^i which minimized the electronic energy. In CI methods, the variational CI energy is variationally optimized with respect to the expansion coefficients $C_{ij...}^{ab...}$. With multi-configuration SCF methods, a limited number of excited determinants is selected, and a CI wavefunction involving these determinants is constructed. The CI energy resulting from this wavefunction is then minimized with respect to changes in both the expansion coefficients for the spinors and the CI coefficients. Clearly this method can be extremely expensive if the number of excited determinants included in the CI space is large. A practical implementation of the Multi-Configuration Dirac-Fock method has been presented by Jensen, Dyall, Saue and Faegri. These wavefunctions are often employed to construct a set of single particle spinors which may then be used to construct a CI wavefunction which employs a larger active space.

SECOND-ORDER PERTURBATION THEORY

In non-relativistic quantum chemistry, one of the most common methods of taking account of electron correlation effects is through second-order MBPT. Its use is common because it is relatively inexpensive as compared to the other available electron correlation methods. When no restrictions are placed upon the spin orbitals which are used to construct the DF wavefunction and the related excited determinants, the application of MBPT generally follows the method suggested by Møller and Plesset (MP), which is based upon the more general Rayleigh-Schrödinger perturbation theory (RSPT). A good overview of MBPT as applied to the solution of the non-relativistic Hamiltonian may be found in one of several elementary texts and reviews. These methods begin with the assumption that the problem of finding eigenfunctions of the Hamiltonian may be divided into a part that is relatively easy to solve, and a part which is more difficult:

$$\widehat{H} = \widehat{H}_0 + \widehat{V}.$$

$\widehat{H}_0$ is known as the unperturbed or zeroth-order Hamiltonian, while $\widehat{V}$ is the perturbation. In relativistic Møller-Plesset perturbation theory, $\widehat{H}_0$ is simply the Dirac-Fock operator. Higher order corrections to the zeroth-order, DF wavefunction, may be constructed via combinations of the other solutions to the zeroth order Hamiltonian, the aforementioned excited determinants. With this partitioning of the Hamiltonian, the zeroth order energy is simply the sum of orbital eigenvalues for the occupied spinors of the DF wavefunction. The first-order correction to the energy only depends upon the zeroth-order wavefunction and simply reproduces the DF energy:

$$E^{(1)} = \left\langle \Psi_{DHF} \mid \hat{V} \mid \Psi_{DHF} \right\rangle$$

$$E_{DHF} - \sum_i \epsilon_i$$

In order to obtain the second order correction to the energy, however, the first order corrections to the wavefunction are required:

$$E^{(2)} = \left\langle \Psi^{(0)} \mid \hat{V} \mid \Psi^{(1)} \right\rangle.$$

The first order correction to Ψ is given by the expansion

$$\Psi^{(1)} = \sum_i^{acc} \sum_a^{vir} C_i^{a(1)} \Psi_i^a + \sum_{ij}^{acc} \sum_{ab}^{vir} C_{ij}^{ab(1)} \Psi_{ij}^{ab} + \ldots$$

Where the expansion coefficients, $C^{(1)}$, are determined via the relation

$$C_{ij}^{ab\ldots(1)} = \frac{\left\langle \Psi^{(0)} \mid \hat{V} \mid \Psi_{ij\ldots}^{ab\ldots} \right\rangle}{\left(E^{(0)} - E_{ij\ldots}^{ab\ldots(0)} \right)}$$

Because of the form of $\hat{V}$, only the $C_{ij}^{ab(1)}$ are non-zero. This determines the second order correction to the energy (and the first order correction to the Dirac-Fock energy)

$$E^{(2)} = -\frac{1}{4} \sum_{ab}^{vir} \sum_{ij}^{acc} \frac{\left| \left\langle ab \mid\mid ij \right\rangle \right|^2}{\epsilon_a + \epsilon_b - \epsilon_i - \epsilon_j}.$$

The DHF MP2 method is much more reasonable computationally than most of the other relativistic, correlated techniques. Evaluation of the DHF MP2 energy requires only a limited transformation of the atomic two-electron integrals. Because only electronic solutions are desired, transformations involving unoccupied, positronic single particle solutions need not be performed. This sort of limitation is enforced for integral transformations preceding any correlated calculation, but in the case of the DHF MP2 energy, the transformation of the first two indexes may be performed over the unoccupied spinors, while transformation of the final pair of indexes is performed over the occupied spinors.

Dyall presents an implementation of the DHF MP2 which further restricts the integrals which require transformation via the implementation of Kramer's' restriction on the spinors and the use of double group symmetry. The low computational cost of the MP2 method is its greatest virtue. Unfortunately, the non-relativistic implementation of this technique is known to be unreliable for a wide variety of systems, and the application of the MP2 method is usually performed only in cases where higher level techniques such as coupled-cluster methods are too costly to perform.

COUPLED-CLUSTER

Coupled cluster (CC) methods begin with an approximate wave function which is somewhat more complicated than those proposed for the CI and MP2 methods. Instead of casting the wavefunction as a simple linear combination of excited determinants, CC methods employ excitation operators, $T_{ij...}^{ab...}$, whose action on the DHF reference wavefunction is given by

$$T_{ij...}^{ab...} \mid \langle ir...jp...q \rangle = t_{ij...}^{ab...} \mid \langle ar...bp...q \rangle$$

where the scalar quantity $t_{ij...}^{ab...}$ is known as an amplitude. These operators are classified according to the number of spinors which their application swaps. The sum of all operators of a particular class forms a single $\hat{T}_n$ operator, for example the sum of all $\hat{T}_i^a$ forms the $\hat{T}_1$ operator. These operators are then employed in an exponential fashion to form a coupled-cluster wavefunction

$$|\Psi^{CC}\rangle = e^{\hat{T}_1+\hat{T}_2+\cdots}|\Psi^{DHF}\rangle.$$

In the case that all the $\hat{T}_n$ operators up to $n = n_e$ are included in the exponential operator, the resultant coupled-cluster wavefunction is equivalent to the FCI wavefunction. Usually, however, only the $\hat{T}_1$ and $\hat{T}_2$ operators are employed to give what is known as the CCSD wavefunction. Left projection by the reference state, a singly excited state vector, and a doubly excited state vector gives three sets of equations which subsequently allow for the iterative solution of t_i^a, t_{ij}^{ab}, and E_{CCSD}. The non-relativistic CCSD method, and the associated CCSD(T) method, which attempts to estimate the effects of the inclusion of the $\hat{T}_3$ operator in the exponential excitation operator, have proved to be extremely successful in providing chemically accurate predictions for a large number of difficult chemical systems.

Practical implemen-tations of the DHF based versions of these methods have been reported by Lindgren as well as Dyall, Lee, and Visscher. Dyall et al. claim that their implementation of CCSD, which employs double group symmetry and Kramer's restricted single particle functions, should employ 32 times the number of floating point operations required by a similar implementation of the equivalent non-relativistic CCSD. The success of the non-relativistic CC methods, the rapid development of modern workstation and supercomputer capabilities, and the existence of highly optimized implementations of these methods in such programme packages as MOLFDIR all indicate that, in the future, relativistic coupled cluster methods will play an important role in the treatment of systems which exhibit significant relativistic and correlation effects.

TWO-COMPONENT METHODS

The four-component wavefunctions of the Dirac-Hartee-Fock method present a practical barrier for the quantum chemical investigator. Methods which decouple the positronic and electronic solutions have been employed with a great deal of success. Though it is not possible to exactly separate the two halves of the DHF solutions for molecular wavefunctions, methods such as the Foldy-Wouthoysen transformation and the related Douglas-Kroll transformation can produce a molecular Hamiltonian which is decoupled to a certain order in the fine structure constant, $\alpha = \frac{1}{c}$. The resultant Hamiltonians, and their two component solutions, restore the traditional, familiar picture of molecular electronic structure where all interactions are solely among electrons and nuclei. This return to the two component picture also brings the advantage of dramatically increased affordability of correlated methods.

THE FOLDY-WOUTHOYSEN TRANSFORMATION

The positronic solutions of the Dirac Hamiltonian and approximate, multi-electron Hamiltonians such as H^{DC} give rise to equations which are more complicated to interpret physically and more difficult to implement computationally. Finite basis set expansion of the model space for such Hamiltonians requires a basis set of at least twice the size of a non-relativistic basis set for similar quality results. Since quantum chemists are used to thinking in terms of purely electronic wavefunctions, the introduction of the positronic terms can serve to rob chemists of some of their intuition for quantum chemical models. Additionally, the existance of the continuum of negative energy eigenstates destroys the chemist's conception of a bound chemical system.

The Foldy-Wouthoysen transformation reduces the coupling between the postitronic and electronic solutions which arise from fully relativistic Hamiltonians. The operators which result from such a transformation can be very complicated, but provide solutions which fit more comfortably into traditional concepts of electronic solutions.

The molecular Dirac-Coulomb hamiltonian contains only one electron operators which couple the large and small components of the wavefunction. Therefore, it is possible to transform $\widehat{H}_{DC}$ to a two-component Hamiltonian by simply transforming the one electron terms of the Hamiltonian. The Dirac single-particle hamiltonian, $\widehat{h}_D$, may be represented as

$$\widehat{h}_D = \begin{pmatrix} \widehat{E}_L & \widehat{O} \\ \widehat{O} & \widehat{E}_S \end{pmatrix}$$

where the even portion of the Hamiltonian, which do not involve interactions between the large and small componenents, are given by

$$E_L = (\phi + mc^2)$$

$$E_S = (\phi - mc^2)$$

and the odd portions, which do describe the interaction between the large and small component, and are given by

$$\widehat{O} = \sigma.\pi$$

A more mathematical definition of even and odd operators is given by the requirement that an even operator, $\widehat{\varepsilon}$, commutes with β, and that the odd operator, $\widehat{O}$, anticommutes with β. It is important to note that any operator, $\widehat{H}$, may be represented as the sum of an even and odd operator, $\widehat{\varepsilon}$ and $\widehat{O}$ where

$$= 1\overline{2(\widehat{H} + \beta\widehat{H}\beta) \text{ and } \widehat{O}}$$

$$= 1\overline{2(\widehat{H} - \beta\widehat{H}\beta)(4.109)}$$

The objective of the Foldy-Wouthoysen transformation is to apply a unitary transform to the relativistic, four-component wave equation in order to decouple the large and small components by eradicating the odd portions of the Hamiltonian to some order in an expansion parameter which will go to zero in the non-relativistic limit. A good form of the unitary transform, $\widehat{U}$, turns out to be

$$\hat{U} = e^{i\hat{S}} = \sum_{n=0}^{\infty} \frac{(i\hat{S})}{n!}$$

where

$$\hat{S} = \frac{-i\beta\hat{O}}{2mc^2} = -i\beta\frac{\alpha.\pi}{2mc}$$

This choice of $\hat{U}$ will result in the transformed electronic wavefunction

$$\phi = e^{i\hat{S}}\psi$$

where ϕ is, effectively, a two component wavefunction, since the contribution of the small component has been removed to some level of approximation. Similarly, the transformed, two-component positronic wavefunction, χ may also be produced via the action of $\hat{U}$ on the four-component positronic wavefunction, ψ, which will effectively remove the large component contribution.

In order to achieve this transformed wavefunction, we consider a somewhat altered version of our original wave equation where the full, four-component wavefunction, ψ, has been expressed as a product of the inverse transformation, $\hat{U}^{-1} = e^{-i\hat{S}}$, and the two component wavefuction, and the entire equation has been pre-multiplied by $\hat{U}$ to give

$$e^{iS}\hat{h}_D e^{-iS}\phi = e^{iS}\frac{\partial}{\partial t}\left(e^{-iS}\phi\right)$$

$$= e^{iS}\left[\frac{\partial}{\partial t}, e^{-iS}\right]\phi + \frac{\partial}{\partial t}\phi.$$

This gives us the definition of the transformed one-particle Hamiltonian,

$$\hat{h}'_d = e^{iS}\hat{h}_D e^{-iS} - e^{iS}\left[\frac{\partial}{\partial t}, e^{-iS}\right]$$

Expanding $\hat{U}$ and $\hat{U}'$ in terms of their power series expansions, and collecting terms of similar order in $\hat{S}$ gives a series of terms which are most efficiently represented by nested commutators

$$\hat{h}'_D = \hat{h}_D - \left[\frac{\partial}{\partial t}, \hat{S}\right] + i\left[\hat{S}, \hat{h}_D - \frac{1}{2}\left[\frac{\partial}{\partial t}, \hat{S}\right]\right]$$
$$+\left(\frac{i^2}{2!}\right)\left[\hat{S}, \left[\hat{S}, \hat{h}_D - \frac{1}{2}\left[\frac{\partial}{\partial t}, \hat{S}\right]\right]\right] + \ldots$$

where the terms which have been omitted are of order $mc^2\alpha^5$. The resultant Hamiltonian, $\hat{h}'_D$, may again be expressed as the sum of even and odd operators, $\hat{\varepsilon}$ and $\hat{O}'$ (and, now, an error term of order $mc^2\alpha^5$) where the new even and odd terms are given by

$$\hat{\varepsilon}' = \hat{\varepsilon} + \frac{1}{2mc^2}\hat{\beta}\hat{O}^2 - \frac{1}{8m^2c^3}\left[\hat{O}, \left[i\frac{\partial}{\partial t} - \hat{\varepsilon}, \hat{O}\right]\right]$$
$$-\frac{1}{8m^3c^\beta}\hat{\beta}\hat{O}^4 + O(mc^2\alpha^5)$$
$$\hat{O}' = \frac{1}{2mc^2}\hat{\beta}\left[i\frac{\partial}{\partial t} - \hat{\varepsilon}, \hat{O}\right]$$

Subsequently, the largest contributing odd term which remains may be removed through the use of the unitary transformation

$$\hat{U}' = e^{\frac{-i}{2mc^2}\hat{\beta}\hat{O}'}$$

to give new even and odd operators which are now higher order in the fine structure constant, and a resudual error of the order $mc^2\alpha^6$. Application of a third transform gives an even operator, $\hat{\varepsilon}'''$ which is correct to order $mc^2\alpha^5$. Truncation at this order represents a very useful stopping point. The third order FW transformed Dirac one-particle Hamiltonian for the electronic solutions is given by

$$\hat{h}_D''' = mc^2 - e\phi + \frac{1}{2m}\pi^2$$

$$+\frac{e\hbar}{m}(\widehat{s}.B)$$
$$-\frac{1}{4m^2c^2}\widehat{s}.(\pi\times E-E\times\pi)$$
$$-\frac{1}{8m^3c^2}\pi^4$$
$$+\frac{e\hbar}{2m^3c^2}(\nabla.E))+O(mc^2\alpha^6).$$

In the absence of an external vector potential, A, the two-component Hamiltonian becomes somewhat simpler:

$$\widehat{h}_D^m = mc^2+e\phi+\frac{1}{2m}$$
$$-\frac{1}{4m^2c^2}\widehat{s}.(\widehat{\pi}\times E-E\times\widehat{\pi})$$
$$-\frac{1}{8m^3c^2}\widehat{\pi}^4$$
$$+\frac{1}{8m^3c^2}(\nabla.E)+O(mc^2\alpha^6).$$

The first three terms of equations may be recognized as the non-relativistic single-particle Hamiltonian. This decoupled, approximate Hamiltonian, which is correct to order $(mc^2\alpha^5)$, may also be obtained via the method of small components. In the method of small components, the single particle wavefunction is replaced by

$$\Psi' = e^{-imc^2/\hbar}\Psi$$

in order to remove the rest mass energy from the Hamiltonian to give the modified single particle equation

$$[\widehat{p}_0+(e/c)-(\alpha.\pi)-(\beta-1)mc]\Psi=0.$$

This relation may be expressed as a pair of coupled differential equations in Ψ_I and Ψ_S, the large and small components.

$$[\widehat{p}_0+(e/c)\phi]\Psi_L=[\sigma.\pi]\Psi_S$$
$$[\widehat{p}_0+(e/c)\phi+2mc]\Psi_S=[\sigma.\pi]\Psi_L$$

Using the second differential equation to obtain an expression for Ψ_S in terms of Ψ_L gives

$$\Psi_S = (\hat{p}_0 + (e/c)\phi + 2mc)^{-1}(\sigma.\pi)\Psi_L$$

Substituting this result into the first differential equation gives

$$[\hat{p}_0 + (e/c)\phi]\Psi_L = (\sigma.\pi)\hat{K}(\sigma.\pi)\Psi_L$$

where

$$\hat{K} = ((\hat{p}_0 + (e/c)\phi)/2mc + 1)^{-1}.$$

Using a rearrangement of

$$(\hat{p}_0 + (e/c)\phi) = 2mc(\hat{K}^{-1} - 1)$$

the action of $\hat{K}$ on Ψ_L is found to be

$$\hat{K}\Psi_L = [1 - \hat{K}(\sigma.\pi)\hat{K}(\sigma.\pi)/(4m^2c^4)]\Psi_L$$

Because it is not possible to solve for $\hat{K}$ analytically, and hence achieve exact separation of Ψ_L and Ψ_S, except in the case of the free particle, it is necessary to utilize the approximate form of $\hat{K}$ given by

$$\hat{K} = [1 - (\sigma.\pi)^2/(4m^2c^2) + O(\alpha^4)]$$

to give the approximate Hamiltonian

$$\hat{H} = -e\phi + (\sigma.\pi)^2/2m$$

$$-(\sigma.\pi)^4/(8m^4c^4)$$

$$+\frac{i\hbar e}{4m^2c^2}(\sigma.\pi)(E.\pi) + O(mc^2\alpha^6)$$

The last term in this Hamiltonian is somewhat disconcerting, since it is not Hermitian. This Hamiltonian, however, was derived for the unnormalized Ψ_L, and the renormalization will neccessarily be dependent upon the magnitude of the coupling of large and small components, and hence on $\hat{K}$. The Hamiltonian for the renormalized wavefunction, Ψ', is given by

$$\hat{H} = -e\phi + (\sigma.\pi)^2/2m$$
$$-(\sigma.\pi)^4/(8m^4c^4)$$
$$+\frac{i\hbar e}{8m^2c^2}[E.\pi, \sigma.\pi] + O(mc^2\alpha^6)$$

This transformed Hamiltonian is identical to the Foldy-Wouthoysen transformed Hamiltonian given in, with the exception that the rest mass energy of the electron, mc^2, has been removed.

PERTURBATIVE RELATIVISTIC TREATMENTS

The approximate Hamiltonian which arises from the Foldy-Wouthoysen transformation and the method of small components is frequently known as the Pauli Hamiltonian, and methods which utilize this Hamiltonian as a starting point are said to stem from the Pauli formulation. In contrast to the Dirac formulation, the Pauli formulation strongly suggests that relativistic effects may be viewed as a perturbation to the non-relativistic case. Most of the terms retained in equations have direct physical interpretations.

The term on the second line of corresponds to the interaction of the electron spin magnetic moment with an external magnetic field, and is generally known as the Zeeman term. The Zeeman term is zeroth order in the fine structure constant, α, and may be obtained directly from classical magnetostatics. The first relativistic correction to this term, roughly second order in the fine structure constant, is given by the fifth line in.

The term on the second line of corresponds to the interaction of the electron spin magnetic moment with the magnetic field produced by the electron's motion within an electric field. In the case that the electric feild is that of the nucleus, this term represents spin-orbit coupling. The fourth term represents the relativistic correction to the kinetic energy term. This term is commonly called the mass-velocity (MV) term.

The final term of gives what is known as the Darwin operator. This operator has no simple classical analogue. In order to get a feel for the physical origin of this term, it is useful to investigate the relativistic velocity operator, $\frac{dr}{dt}$. In non-relativistic quantum mechanics, the velocity operator is given simply by the classically intuitive $\frac{\hat{p}}{m}$. The relativistic analog, on the other hand, is given by

$$\frac{dr}{dt} = \frac{i}{\hbar}[\hat{H}, r]$$

$$= \frac{i}{\hbar}[c\alpha.\hat{p}, r] = c\alpha$$

Since the eigenvalues of any single component of $c\alpha$ may only take the value $\pm c$, this gives the somwhat disturbing result that each component of the velocity of an electron has a magnitude c. Further insight may be divined by inspection of the time derivative of $c\alpha$. Such an analysis reveals that even slow moving electrons appear to undergo an extremely small amplitude, highly oscillatory motion about its average postition which has been termed *Zitterbewegung*, or "trembling motion". The Darwin term represents the first order correction to the "average position" model of non-relativistic quantum mechanics.

One of the most popular methods which employ the Pauli Hamiltonian is known as the MVD or mass-velocity plus Darwin method. In the MVD method, only the *j*-independent terms of the Pauli Hamiltonian are retained, to give

$$\hat{H}_{MVD} = \frac{\hat{p}^2}{2m} + e\phi - \frac{\hat{p}^4}{8c^2} + \frac{\pi Z}{2c^2}\delta(r)$$

Once a converged HF wavefunction has been obtained, the MVD Hamiltonian may be employed in order to obtain a estimate of the first order relativistic correction to the electronic energy. Unfortunately, because evaluation of the Darwin term leads to singularities in the region of the nucleus, and the fact that the contributions from the mass-velocity term diverge with higher order applications, the perturbative expansion of the energy may not be extended beyond first-order for molecular systems. These effects are a result of the finite basis set expansion of the electronic wavefunction. Because only the first-order correction to the electronic energy may be obtained, no estimate of the relativistic effects on the wavefunction may be obtained.

For atomic systems, the singularities associated with the MVD Hamiltonian may be avoided by solving for the atomic radial wavefunctions numerically. Cowan and Griffin first proposed the application of the MVD Hamiltonian to atomic systems in 1976. By incorporating the mass-velocity and Darwin terms in the variational optimization of the radial wavefunctions in what they called the HFR method, Cowan and Griffin were able to produce atomic spin orbitals which exhibited orbital energies and $\left\langle r^n \right\rangle$ values which closely matched those of the *l*-averaged atomic orbitals produced by the DHF method for atoms as heavy as uranium.

THE DOUGLAS-KROLL TRANSFORMATION

Another method of decoupling the large and small components of the DHF wavefunctions is given by the Douglas-Kroll method. First presented by Douglas and Kroll in 1974, this method begins with a first order Foldy-Wouthoysen transformation performed in momentum space. The momentum space Hamiltonian is given by

$$\hat{h}_D = \alpha.p + \beta m + \hat{V}_x$$

where the action of the external potential, $\hat{V}_x$, on a momentum space state vector, $\phi(p)$, is given by the expression

$$\hat{V}_x \phi(p) = \int d^3p' V(p,p')\phi(p').$$

The unitary operator used in the first-order transformation is constructed using those free-particle momentum-space eigensolutions of the Dirac Hamiltonian associated with the positive energy eigenvalues

$$\hat{E}_p = (p^2 + m^2)^{1/2}.$$

The resultant unitary transformation operator is given by

$$\hat{U} = \left(\frac{\hat{E}_p + m}{2\hat{E}_p}\right)^{1/2} \left[1 + \beta\left(\frac{\alpha.p}{\hat{E}_p + m}\right)\right]$$

When this unitary operator is applied to the single particle Hamiltonian, the result is given by

$$\hat{U}\hat{h}_D\hat{U}' = \beta\hat{E}_p + \hat{\varepsilon} + \hat{O} = \hat{h}'_D$$

with $\hat{\varepsilon}$ and $\hat{O}$ given by

$$\hat{\varepsilon} = \left(\frac{\hat{E}_p + m}{2\hat{E}_p}\right)^{1/2} \left(\hat{V}_x + \frac{\alpha.p}{\hat{E}_p + m}\hat{V}_x \frac{\alpha.p}{\hat{E}_p + m}\right)\left(\frac{\hat{E}_p + m}{2\hat{E}_p}\right)^{1/2}$$

$$\hat{O} = \beta\left(\frac{\hat{E}_p + m}{2\hat{E}_p}\right)^{1/2} \left[\frac{\alpha.p}{2\hat{E}_p + m}, \hat{V}_x\right].$$

Use of the definitions

$$\hat{A}_p = \left(\frac{\hat{E}_p + m}{2\hat{E}_p}\right)^{1/2}$$

$$\hat{R}_p = \frac{\alpha.p}{\hat{E}_p + m}$$

provides more compact expressions for the even and odd operators

$$\hat{\varepsilon} = \hat{A}_p(\hat{V}_x + \hat{R}_p\hat{V}\hat{R}_p)A$$

$$\hat{O} = \beta\hat{A}_p[\hat{R}_p, \hat{V}_x]$$

From this point, Douglas and Kroll proposed the use of a second unitary transformation, of the form

$$\hat{U}' = (1 + \hat{W}_1^2)^{1/2} + \hat{W}_1$$

where $\hat{W}_1$ is anti-Hermitian, and is selected such that, in the resulting transformed Hamiltonian

$$\hat{U}\hat{h}'_D\hat{U}' = \beta\hat{E}_p - [\beta\hat{E}_p, \hat{W}_1] + \hat{\varepsilon} + \hat{O}$$
$$+\frac{1}{2}[\beta\hat{E}_p, \hat{W}_1^2]_+ - \hat{W}_1\beta\hat{E}_p\hat{W}_1$$
$$+[\hat{W}_1, \hat{O}] + [\hat{W}_1, \hat{\varepsilon} + \ldots$$

the $\hat{O}$ term is exactly canceled by the $[\beta\hat{E}_p, \hat{W}_1]$ term. This condition is fufilled if $\hat{W}_1$ obeys the equation

$$[\hat{W}_1, \hat{E}_p]_+ = \beta\hat{O}.$$

Because the operator $\hat{O}$ is linear in $\hat{V}_x$, an integral operator, $\hat{W}_1$ must be expressed in terms of its kernel

$$\hat{W}_1(p, p') = \beta\frac{\hat{O}}{E_{p'} + \hat{E}_p}.$$

Using the functional form of $\hat{O}$, a more explicit expression for the anti-Hermitian operator, $\hat{W}_1(p, p')$ may be obtained

$$\hat{W}_1(p, p') = A_{p'}\hat{A}_p(R_p - R'_p)\frac{V_x(p, p')}{E_{p'} + \hat{E}_p}$$

This transformation removes all coupling of the upper and lower halves of the transformed wavefunction to first order in V_x. Douglas and Kroll suggested that further separation, to arbitrary orders in the external potential were possible through the repeated application of the *n* order transformation

$$\hat{U}_n = (1 + \hat{W}_n)^{\frac{1}{2}}\hat{W}_n$$

where $\hat{W}_n$ is, again, an anti-Hermetian operator linear in the external potential. Separation to second order in the external potential represents a practical stopping point, since the higher order transformations become extremely complicated. Furthermore, higher order contributions should only be important for the proper treatment of the lowest lying electrons, and so should not substantially affect the description of the valence electronic structure.

The large component block of the single-particle Hamiltonian which has been decoupled to second order in V_x is given by

$$\hat{h}_D^n = \hat{E}_p + \hat{\varepsilon}_1 + 1/2\left(\left[\hat{E}_p, \hat{W}_1^2\right]_+ + \hat{W}_1\hat{E}_p\hat{W}_1\right).$$

The resultant single-particle Hamiltonian may be employed used in conjunction with the coulomb inter-electron repulsion operator to define a many-electron hamiltonian:

$$\hat{H} = \sum_i^{N_e} \hat{h}_D^n + 1/2\sum_{ij}^{N-e} \frac{1}{r_{ij}}.$$

This Hamiltonian is frequently known as the no-pair Douglas-Kroll Hamiltonian, and it has been extensively discussed and tested by Hess and coworkers. Molecular wavefunctions may be obtained with the no-pair DK Hamiltonian in much the same fashion that solutions are obtained in the non-relativistic case, and investigations of systems containing heavy atoms by means of DK based electronic structure methods have given very encouraging comparison to fully relativistic results.

5 RELATIVISTIC EFFECTIVE CORE POTENTIALS

One of the most fundamental assumptions in chemistry is that low lying core electrons are relatively inert, and are not perturbed by a molecular environment. This assumption is supported by the chemical similarity of elements in the same column of the periodic table. Most of the important chemical properties of atoms and molecules are determined by the interaction of their valence electrons with the valence electrons of other atomic or molecular species.

For molecules containing heavy atoms, large computational savings may be realised if a particular mathematical form for the atomic orbitals is assumed, *a priori*, so that the total number of parameters which must be optimized in, say, the computation of a Hartree-Fock wavefunction may be reduced.

A more extensive savings may be achieved if, instead of simply assuming a particular form for the core orbitals, all terms describing the interaction of the electrons in these orbitals with each other and those outside the core region are simply replaced by a scalar "effective potential". This provides the additional benefit of reducing the basis set size requirements, since no basis functions are now needed to explicitly describe the core orbitals. This has the effect of drastically reducing the number of two-electron quantities, such as electron repulsion integrals, which must be considered.

In correlated electronic structure calculations, the molecular orbitals which correspond to atomic core orbitals are often excluded from the active orbital space. Even when such orbitals are included in the active space, the resultant contributions are typically small for most chemical properties. The use of effective core orbitals in these methods ideally should not significantly affect the quality of the resulting molecular property predictions. Because correlated methods scale higher than Hartree-Fock with respect to basis set size, the decreased basis set size resulting from the use of effective potentials will introduce in an even more dramatic computational savings over all-electron methods.

EFFECTIVE CORE POTENTIALS

The objective of the effective core potential method is to construct potentials which are solely dependent upon the coordinates of the valence electrons, but take into account the influence of the inert, core electrons. The all electron Hartree-Fock model for an n-electron system begins with an antisymmetrized Hartree product wavefunction of the form

$$\Psi_{HF} = A(\phi_1\phi_2 \cdots \phi_n)$$

where A is the antisymmetrizer operator and $\{\phi_i\}$ are the single-particle eigenfunctions of the Fock operator

$$\hat{F} = -\frac{\nabla_r^2}{2} - \frac{Z}{r} + \frac{\ell(\ell+1)}{2r^2} + \sum_{j=1}^{n}(\hat{J}_j - \hat{K}_j)$$

where the action of the operators $\hat{J}_j$ and $\hat{K}_j$ is defined by

$$\hat{J}_j\phi_i(r_1)=\left(\int dr_2\phi_j^*(r_2)\frac{1}{r_{12}}\phi_i(r_2)\right)\phi_i(r_1)$$

$$\hat{K}_j\phi_i(r_1)=\left(\int dr_2\phi_j^*(r_2)\frac{1}{r_{12}}\phi_i(r_2)\right)\phi_j(r_1)$$

If we now divide the orbitals into a group of N_C core orbitals and N_v valence orbitals, we may re-express the Fock operator as

$$\hat{F}=-\frac{\nabla^2}{2}-\frac{Z}{r}+\frac{\ell(\ell+1)}{2r^2}+\hat{V}^{core}+\hat{V}^{val}$$

where

$$\hat{V}^{core}=\sum_a^{N_C}(\hat{J}_a-\hat{K}_a) \text{ and } \hat{V}^{val}=\sum_i^{N_v}(\hat{J}_i-\hat{K}_i).$$

In order to make the computational task of solving the Fock equations for the molecular case more affordable, it would be useful to develop an equation analogous to the Fock equation which does not contain non-local terms which depend upon the core orbitals. Such an equation would take the form

$$\left(-\frac{1}{2}\nabla_r^2-\frac{Z_{eff}}{r}+\frac{l(l+1)}{2r^2}+\hat{V}^{val}+\hat{V}^{eff}\right)\phi_i=\epsilon_I\,\phi_i$$

where the full coulomb and exchange terms of $\hat{V}^{core}$ have been replaced by an effective core potential (ECP), $\hat{V}^{eff}$, a local potential, and the nuclear charge, Z, has been replaced by $\hat{V}_{eff}=Z-N_C$, where N_C is the number of core electrons. The form of this effective potential is typically derived from numerical Hartree-Fock atomic solutions, in accordance with the frozen core approximation.

Several methods have been suggested for obtaining $\hat{V}^{eff}$. If we consider equation for a valence atomic orbital of angular momentum l, ϕ_i^l, it is possible to obtain an analytic form for $\hat{V}_l^{eff}$, the local potential which will exactly reproduce ϕ_i^t

$$\hat{V}_l^{eff} = \in_l + \frac{Z_{eff}}{r} - \frac{l(l+1)}{2r^2} + \frac{\left(\frac{1}{2}\nabla_r^2 - V_{val}^{'}\right)\phi_i^t}{\phi_i^l}$$

where the l subscript acknowledges the unique form of $\hat{V}^{eff}$ for each value of valence orbital angular momentum l.

Unfortunately, this expression for $\hat{V}_l^{eff}$ is only valid for ϕ_i, since the last term in equation is singular when $\phi_i = 0$. To circumvent this problem, the full atomic valence orbitals, $\{\phi_i\}$, may be replaced by approximate pseudo-orbitals, $\{\chi_i\}$, which are nodeless.

Ideally we would like to obtain a $\hat{V}_l^{eff}$ which will produce atomic valence pseudo-orbitals which are as close to the original orbitals, $\{\phi_i\}$, as possible. The use of approximate valence orbitals brings a new definition of $\hat{V}^{eff}$, since the valence orbitals now satisfy the equation

$$\left[-\frac{1}{2}\nabla_r^2 - \frac{Z}{r} + \frac{l(l+1)}{2r^2} + V_{val}^{'} + V_l^{eff}\right]\chi_i^l = \in_i^l \chi_i^l.$$

Because $V_{val}^{'}$ will differ from V_{val}, the ECP must represent not only the core-core and core-valence interactions, but also the parts of the valence-valence interaction which were lost in the switch from ϕ_i to χ_i.

If the difference between $V_{val}^{'}$ and V_{val} is large, then the ECP, which will contain valence-valence interaction terms for the atomic case, will bias the potential in the molecular valence region towards the atomic case.

One popular method of forming the nodeless χ_i takes a linear combination of the appropriate full electron valence orbital, ϕ_v, with all of the full electron core orbitals, χ_c, to give

$$\chi_v = a_v\phi_v + \sum_c a_c\phi_c,$$

where the parameters u_v and u_c are selected by minimizing the kinetic energy of χ_v. Because of the imposition of the normalization condition upon χ_v, a_v will always be less than one, which means that χ_v will be less than ϕ_v outside of the core region, where the ϕ_c's will have little to no contribution. This could potentially introduce a substantial difference in V_{val} and V'_{val} and thereby introduce large valence-valence terms to the ECP.

One χ_I selection scheme which gets around the problem of valence-valence contributions to the ECP is the technique of selecting shape consistent pseudo-orbitals. This method requires that χ_i exactly match the original valence atomic orbital, ϕ_i, for $r \geq R_{\text{max}}$, where R_{max} is the radius at which ϕ_i experiences its outermost maximum. Inside this radius, the wavefunction is represented by a function of the form

$$\chi_i = \sum_{m=0}^{4} c_m r^{m+N}, r < R_{\text{max}}$$

where $N = 1 + 2$, in most cases. The coefficients, c_m, are restricted by the requirement that the zeroth through third derivatives of χ_i and ϕ_i are equal, and that χ_i is normalized. The resultant χ_i are generally known as shape-consistent pseudo-orbitals.

By solving for V_I^{eff} we obtain

$$V_I^{eff} = \epsilon_i^I + \frac{Z_{eff}}{r} - \frac{l(l+1)}{2r^2} + \frac{\left(\frac{1}{2}\nabla_r^2 - V'_{val}\right)\chi_i^t}{\chi_i^I}.$$

The total V_{eff} for an atom, then, is written

$$V_{eff} = \sum_I V_I^{eff} \sum_{m=-l}^{t} |lm\rangle\langle lm|$$

so that the orthonormality of the spherical harmonics, $|lm\rangle$, pairs each atomic wavefunction with the proper V_I^{eff} term. The radial wavefunction of the valence orbitals is required to be orthogonal to the core orbitals of the same angular momentum as that valence orbital.

If an atomic core contains orbitals with angular momenta up to a certain value l_{max} then valence orbitals of a greater orbital angular momentum than l_{max} do not have any such orthogonality constraints on their radial wavefunctions. V_I^{eff} for $\chi_I(l > l_{max})$, then, will be largely independent of 1, since the only 1 dependent terms which should be represented are the core-valence exchange terms, which should not be very large. It is convenient, then, to approximate as

$$V_{eff} = V_{I_{max}} + \sum_{I}^{I_{max}-1} V_I^{eff} - V_{I_{max}} \sum_{I} V_I^{eff} \sum_{m=-I}^{I} |lm\rangle\langle lm|.$$

Once the nodeless pseudo-orbitals have been selected, it is possible to simply solve numerically. Once this has been done, a parameterized analytic form may be fit to the numerical data. One common form is given by

$$V_I(r) = \sum_i A_i \frac{e^{-\alpha_i r^2}}{r_i^{N_i}}.$$

Once the ECP has been been modeled by some set of functions, then is complete and may be implemented for a chosen set of valence orbitals.

RELATIVISTIC EFFECTIVE CORE POTENTIALS

Relativistic corrections to the Hartree-Fock model tend to be largest in the region immediately surrounding the nucleus. This is because of the fact that the kinetic energy of the electrons is greatest in this region. Therefore, it is reasonable to assume that the electrons in the innermost core orbitals will experience the effects of relativity much more directly than will the higher lying valence orbitals, which tend to have much less probability of existing close to the nucleus.

The interaction of the valence electrons, then, with the nuclei and other electrons should be relatively well described by the non-relativistic Hamiltonian. Although the valence orbitals do not experience large direct relativistic effects, the changes in the core orbitals will have direct consequences on the nuclear shielding and orthogonality constraints which the valence orbitals experience. These are generally known as secondary effects. Since the behaviour of valence electrons is the primary concern for quantum chemists, and the deeper core electrons of very heavy atoms should not change significantly in a molecular environment, an ECP method which incorporates the direct relativistic effects experienced by the core orbitals is of great practical interest.

Relativistic effective core potentials (RECP) may be obtained in a manner directly analogous to the non-relativistic ECP's. The starting point for RECP's is the atomic Dirac-Coulomb-Fock equations. The valence solutions, $\{\phi_i\}$, are four-component spinors, but the large radial component generally accounts for over 99% of the electronic density. Therefore, the re-normalized large components are used as the starting points for the relativistic pseudo-orbitals, χ_i.

The methods used to derive the relativistic pseudo-orbitals are exactly the same as the methods used in the non-relativistic case. Once $\{\chi_i\}$ have been determined, the RECP's are obtained via a relation similar to equation. The DHF eigenstates, however, are now eigenstates of the total angular momentum operator, $\hat{j} = \hat{l} + \hat{s}$, and so the resultant RECP's are no longer uniquely defined for a particular 1 value. Instead, the RECP's are dependent on l and j and are based on the relation

$$\left[-\frac{1}{2}\nabla_r^2 - \frac{Z}{r} + \frac{l(l+1)}{2r^2} + \hat{V}_{val}' + \hat{V}_{I,j}^{eff}\right]\chi_i^{I,j} = \epsilon_i^{I,j}\,\chi_i^{I,j}.$$

Since the RECP's are typically used within the framework of non-relativistic quantum chemistry, where single particle states are eigenfunctions of $\hat{l}$ and $\hat{S}$ rather than $\hat{j}$, it is necessary to construct a V^{eff} which is only 1 dependent. This may be accomplished by statistically averaging over all of the appropriate j-dependent RECP's which are associated with a particular 1 value to give what are known as 1-averaged RECP's (AREP).

$$V_I^{AREP}(r)=(2l+1)^{-1}\left[lV^{REP}_{I,j=1-\frac{1}{2}}(r)+(l+1)V^{REP}_{I,j=I+\frac{1}{2}}(r)\right].$$

The total V^{AREP} may be given by

$$V^{AREP}=\sum_I V_I^{AREP}(r)\sum_{m=-I}^{I}|lm\rangle\langle lm|,$$

completely analogous to the equation for V^{eff}.

Alternatively independent RECP's may be derived by starting with l-averaged pseudo orbitals. Such orbitals may be obtained by statistically averaging the large component of the valence spin orbitals, thereby producing l averaged atomic orbitals which are then utilized to construct nodeless pseudo-orbitals

$$\phi_I=(2l+1)^{-1}(l\phi_{I,j=I-1/2}+(l+1)\phi_{I,j=I+1/2})$$

The *l*-averaging may also take place in an analogous fashion after the *j*-dependent pseudo-orbitals, $\chi_{I,j=I\pm1/2}$, have been formed.

When RECP's are utilized to form valence *LS* eigenstates, the nuclear shielding and core shrinking effects are taken into account, but the third major affect of relativity on the valence orbitals, spin-orbit splitting, is not taken into account. The portion of the original RECP's which was dropped in the process of statistical averaging over *l*-states may be utilized to define a spin-orbit interaction term

$$\hat{H}^{SO}=V^{REP}-V^{AREP}.$$

This spin-orbit interaction operator, $\hat{H}^{SO}$, may be expressed as

$$\hat{H}_{SO}=\sum_{I=1}^{I_{\max}}\left(\hat{V}^{REP}_{I,j=\left(I+\frac{1}{2}\right)}-\hat{V}^{REP}_{I,j=I-\frac{1}{2}}\right)$$

$$\left[\left(\frac{l+1}{2l+1}\right)\sum_{m=I-1/2}^{I+1/2} |l,j=l+1/2,m_j\rangle\langle l,j=l+1/2,m_j| -\right.$$

$$\left.-\left(\frac{l}{2l+1}\right)\sum_{m=-I+1/2}^{I-1/2} |l,j=l-1/2,m_j\rangle\langle l,j=l-1/2,m_j|\right]$$

Since this correction is based solely upon the RECP's, only one-electron integrals are required for its evaluation. This correction term was first presented in the context of CI calculations, but may be employed to estimate the extent of spin-orbit coupling of two HF wavefunctions.

AB INITIO MODEL POTENTIALS

One of the drawbacks of ECPs is that the valence orbitals which are employed do not exhibit the radial nodal structure which naturally arises when all-electron methods are employed. The *ab initio* model potential (AIMP) method, first suggested by Schwartz and Switalski and more extensively developed by Bonifacic and Huzinaga, attempts to ameliorate this unsavory feature of ECPs. The fundamental idea of the AIMP method is that a projection operator, $\hat{p}$, constructed from an appropriate set of atomic core orbitals, can be appended to the Hamiltonian in order to project the atomic core solutions out of the valence wavefunction. In their 1973 work, Bonifacic and Huzinaga present the modified single-particle atomic Hamiltonian

$$\hat{h}_i = -\frac{1}{2}\nabla_i^2 - \frac{(Z-n_c)}{r_i} - \sum_j^{n_c} A_j \frac{e^{-\alpha_j r_i^2}}{r_i} + \sum_j^{n_b} B_j \Omega_j$$

where n_a is the number of expansion terms for the effective potential, n_b is the number of frozen frozen atomic shells, A_j, α_j, and B_j are expansion parameters, and Ω_j is the projection operator for a particular frozen shell of angular momentum l and principle quantum number n. This single-particle Hamiltonian is then employed in conjunction with the Coulomb operator to produce an atomic, multi-electron Hamiltonian,

$$\hat{H} = \sum_{i}^{n_v} \hat{h}_i + \sum_{i<j} \frac{1}{r_{ij}}$$

The various parameters of the single-particle Hamiltonian are then optimized to reproduce the desired valence atomic orbital energies and orbital shapes when the Hamiltonian is applied to the atomic system.

The more recent implementation due to Barandiaran, Seijo, and Huzinaga employ the atomic Cowan-Griffin-Hartree-Fock (CGHF) orbitals in the construction of their projection operators, and in their optimization of the various parameters which are a part of the one-electron Hamiltonian. Because the CGHF method deals entirely with scalar-type wavefunctions, the resultant single-particle functions are not **j**-dependent, and therefore may be used in the construction of such scalar operators.

In order to optimize the various parameters associated with the Hamiltonian, an all-electron basis set must be used not only in the construction of the CGHF wavefunction, but for the atomic AIMP optimization process. Once the form of the model potential has been optimized, the valence basis set is truncated by removing the Gaussian primitives with the highest radial exponents. The truncation is performed to such an extent as to assure that the outer orbital shape and orbital eigenvalues of the valence orbitals may still be reproduced by application of the AIMP method to the atomic system. The resultant truncated basis set can often be reduced one half the size of the original, all-electron basis without significantly compromising the quality of the predictions.

DEFINING THE CORE

The definition of the core-valence division in the construction of ECP's is usually done on the basis of chemical intuition. For the first and second row atoms, this typically means that the *ns* and *np* orbitals make up the valence space. For transition metals, the valence space will include the $(n-1)d$ orbitals. For group IIIA and higher, however, the -orbitals are usually assigned to the core space, since the shell is full and they are considered chemically inert. These standard definitions of the core-valence division are appealing from the point of view that they offer a simple, chemically appealing model of the valence space, and they lead to a drastic reduction in the total number of variables required to describe a full molecular wavefunction, especially for molecules containing very large, main group atoms.

There are many instances, however, where the standard definitions of the core-valence division are inadequate. One problem arises in the case of transition metals, where the $(n-1)d$ -orbitals are considered valence but the $(n-1)s$ and $(n-1)p$ -orbitals are considered part of the core. Because the *d, s,* and *p* orbitals in the same principal quantum shell typically have a significant radial overlap, the exchange terms between them can be significant, and they play an important role in the description of the valence *d* orbitals. The local interaction terms of the effective potential terms, therefore, may be inadequate. To rectify this, it is necessary to include the entire $(n-1)$ shell in the valence space. This has the added benefit of producing *ns* and *np* orbitals with at least one radial node. This altered definition of the core space has proven to give more accurate results for a variety of molecular systems.

A related problem arises for the group IIIA and higher main-group elements. The radial extent of the full $(n-1)d$ shell tends to be more diffuse than that of the $(n-1)s$ and $(n-1)p$ shells. As a result, the *d* orbitals share a significant spatial overlap with the valence orbitals, and the local approximation of the full coulomb and exchange integrals again becomes a poor estimate.

This relationship is accentuated in the heavier elements where the valence *s* and *p* orbitals experience relativistic contraction, while the *d* orbitals experience relativistic expansion (see Chap. 3 for a discussion of the effects of relativity on atomic electronic structure). In these cases, the valence space may be augmented by only the $(n-1)d$ functions, or the total (n – 1) shell, depending on the importance of *s-d* and *s-p* exchange for a proper description of the molecular system in question.

A priori definition of the core space, then, is seen to be a difficult task. Some clues may be taken from the atomic, all-electron solutions. The spatial extent of supposed core electrons may be compared to the spatial extent of valence electrons, and the dependency of the radial functions of the higher-lying core orbitals on the atomic state may both be used as core selection criteria. Ultimately, however, the results of several levels of approximation must be compared to investigate the stability of the predictions to an increase in the core space for each system which is investigated.

RELATIVISTIC EFFECTS IN ATOMS

The importance of "relativistic effects" in properly describing the electronic structure of molecules containing heavy elements is often stressed in the literature. A logical definition of these effects might be:. "The differences in molecular property predictions made by the relativistic Hamiltonian and the non-relativistic Hamiltonian". However, this definition is impractical for several reasons. The first problem with this definition is that it requires the existence of an exact relativistic Hamiltonian. This may be resolved by choosing a single, well-accepted approximate Hamiltonian such as the Dirac-Coulomb or Dirac-Breit Hamiltonian as *the* relativistic Hamiltonian. Another problem stems from the extensive discussion of "correlation effects" in the quantum chemical literature.

This term is used to describe the difference between properties predicted by the Hartree-Fock wavefunctions and the properties predicted by higher level, correlated wavefunctions. In order to study relativistic effects independently of correlation effects, it is useful to define relativistic effects as the differences between the predictions of the Hartree-Fock (HF) and Dirac-Hartree-Fock (DHF) methods. In this way, both relativistic effects and correlation effects may be viewed as corrections to the non-relativistic HF model.

It is important to note, however that the two types of effects should not be regarded as separate, additive effects, and that significant interplay between the two should be expected in cases where relativistic and correlation effects are significant.

Unfortunately, there are other approximations inherent in the HF and DHF methods structure methods which stand in the way of isolating those differences which are purely due to the introduction of relativity. The properties predicted by a particular molecular wavefunction are often highly dependent on the quality of the finite basis set in which the wavefunction is expanded. This is especially true in the case of the DHF method, where the reliability of the predictions are not only dependent upon the size and flexibility of the basis set, but also upon the proper balance of large and small component basis sets. The extraordinarily different basis set requirements of the two methods make it difficult to determine what would constitute an equal treatment of the two.

For atomic systems, however, the issue of basis set effects may be avoided by solving the one dimensional DHF and HF radial equations numerically. This makes atomic systems a good starting point for the investigation of isolated relativistic effects. Furthermore, characterization of the differences in such properties as orbital shape and extent, and orbital energies for isolated atomic systems can help to identify periodic trends in relativistic effects without the perturbation of a molecular environment. Models established in the high symmetry of the central force field of the atom may then lend structure to descriptions of the valence orbital effects in molecules, where significant orbital mixing may be encouraged by the lower symmetry molecular environment.

When molecular systems are too large to be treated with the relatively rigorous, DHF-based four-component methods, more approximate two-component methods such as relativistic perturbation theory are utilized in an attempt to include the most important relativistic effects. In these cases, it is essential to have an estimate of the magnitude of the difference between the relativistic and non-relativistic pictures in order to evaluate the validity of perturbative schemes.

DIRECT, INDIRECT, AND SPIN-ORBIT EFFECTS

A crude estimate of the importance of relativity in characterizing the atomic orbital energies and spatial extent may be achieved by invoking the Bohr model of the hydrogenic atom of charge Z. In this classical picture, an electron in an orbit designated by principal quantum number n would travel at a radius r_n and a velocity v_n with a total energy E_n given (in a.u.) by

$$r_n = \frac{n^2}{mZ}$$

$$v_n = Z/n$$

$$E_n = -\frac{mZ^2}{2n^2}$$

where m has been retained intentionally. Since the electron is participating in accelerated motion, the equations of special relativity cannot apply rigorously. However, it is possible to achieve an estimate of the changes which a relativistic Hamiltonian would introduce for this system by replacing m with the velocity-dependent mass

$$m = \frac{m_0}{\sqrt{1 - v^2/c^2}} = \frac{m_0}{\sqrt{1 - Z^2/n^2c^2}}.$$

This relationship is illustrated in Figure.

The manifestation of this behaviour is typically classified as a direct relativistic effect. Direct effects are most obvious for the *s* and *p* type orbitals since these orbitals tend to have a greater density near the nucleus, and hence experience a greater fraction of the full nuclear charge.

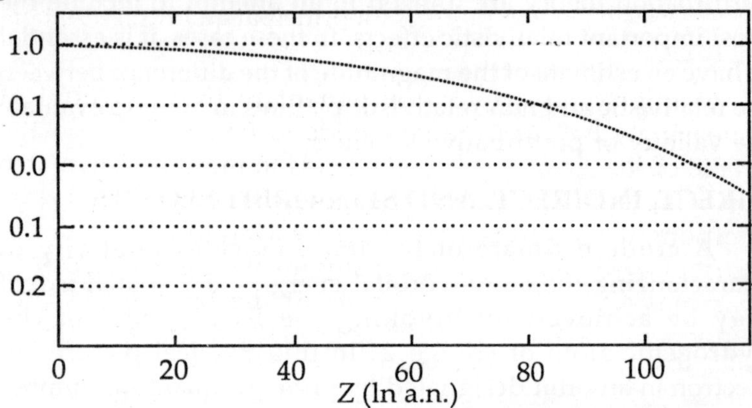

Fig. Ratio of Relativistic Bohr Orbit to Non-Relativistic Bohr Orbit

The orbital shrinking experienced by *s* and *p* orbitals will, in turn, have an affect on the higher angular momentum *d* and *f* orbitals. The greater density of the *s* and *p* orbitals in the nuclear vicinity effectively crowds out the *d* and *f* orbitals through orbital orthogonality and shielding effects. The resultant expansion of the *d* and *f* orbitals and the associated shift of the orbital energies is generally classified as an indirect relativistic effect. In atomic systems where this orbital shrinking and expansion is highly pronounced, non-relativistic methods will obviously be inadequate.

Another litmus test of the necessity of a relativistic treatment is the contribution of the small component to the total electron density predicted by the DHF method for a particular atomic orbital. The pattern of electron density and orthogonality relationships associated with a four-component wavefunction with appreciable large and small components are difficult to reproduce with a two-component wavefunction. The small component of the wavefunction tends to be largest very close to the nucleus and therefore plays a substantial role in the screening of nuclear charge for the higher energy orbitals. The contribution of the small component is usually greatest for orbitals with low principal quantum numbers. This is illustrated by Figure which depicts the large and small components of the radial wavefunction for the **s**-like orbitals ($j = 1/2, k = -1, l = 0$) of radon with principal quantum numbers $n = 1, 2, 3$.

Similarly, the small component to the 1s orbital of increasingly lighter elements tends to be smaller, and the profile of the small components tends to more closely match that of the large component. Therefore, the renormalized large component should give a closer estimate of the radial distribution of the electron density for the lighter elements. This may be observed the plots of the large and small components of the 1s-like DHF orbitals of Rn, Xe, and Kr depicted.

Spin-orbit (S – O) coupling is one of the most manifest of relativistic effects in chemistry. A simple picture of S – O coupling may be obtained by returning to the Bohr atom and employing the classical expression for the interaction of the electron's spin and the magnetic field which results from its orbital motion. An electron orbiting a nucleus may be said to experience the classical magnetic field

$$B = (E \times v / c)$$

where E for a central potential $V(r) = \frac{z}{r}$ is given by

$$E = \frac{1}{r}\frac{\partial V}{\partial r} r$$

The interaction energy, E_{s-o}, of the intrinsic magnetic moment of the electron with this magnetic field is given by

$$\mu B = \left[-\left(\frac{e\hbar}{2mc} \right) \sigma \right] . E \times p$$

$$= \frac{e\hbar}{2mc} \sigma . \left[\left(\frac{1}{r}\frac{\partial V}{\partial r} r \right) \times p \right]$$

$$= \frac{e}{mc}\frac{1}{r}\frac{\partial V}{\partial r} l.s$$

where the definition $s = \frac{\hbar}{2}\sigma$ has been employed. This classical interaction term turns out to be too large by a factor of 2. The reason for this is directly related to the fact that our picture has assumed that the central nucleus is stationary. This classical picture provides a good model for the sorts of interactions which will give rise to the atomic fine structure. The Dirac-Coulomb Hamiltonian accounts for this sort of spin-orbit coupling in multi-electron atoms, but the interaction of the spin of an electron with the orbital motion of another electron and the interaction of one electron spin with that of another are not treated by the DC Hamiltonian. In order to take these effects into account, the Breit interaction term must be added. Fortunately, such interaction terms are typically very small, and do not have a large effect on atomic electronic structure, but they must be included for even qualitative predictions of the atomic spectra of the heaviest atoms.

PERIODIC TRENDS IN THE MAIN GROUP

Traditional, non-relativistic quantum chemistry has been extremely successful in providing high quality, reliable predictions of the properties of ground state molecules containing first and second row elements. The transition metals pose a variety problems for quantum chemical investigators. While relativistic effects can be chemically significant for molecular species containing even third row transition metals, property predictions for even the ground state are often complicated by the presence of many close lying states. This makes it difficult to discuss relativistic and correlation effects separately for species containing these atoms. Atoms from groups IIIA-VIII-A, on the other hand, tend to have less complicated electronic structure, and, in general, their predicted electronic structure is less dependent upon the level of electron correlation employed. For these reasons, it is instructive to investigate the periodic trends of relativistic effects for atoms in this region of the period table.

One useful gauge of the magnitude of the well-established relativistic *s* and *p* orbital shrinkage and *d* orbital expansion for the main group elements is given by the difference in the expectation values of R between the HF and DHF models for these elements. These differences are illustrated in Figure 3.1 for the noble gas elements, while Figures 3.2, 3.3 and 3.4 display these values for the main group elements of the third, fourth, and fifth rows of the periodic table, respectively. The reported values, $\Delta_{I,j}$, may be expressed as

$$\Delta_{I,j} = \left\langle \psi_I^{HF} \mid R \mid \psi_I^{HF} \right\rangle - \left\langle \psi_{I,j=I\pm1/2}^{DHF} \mid R \mid \psi_{I,j=I\pm1/2}^{DHF} \right\rangle.$$

Figure clearly depicts the onset of chemically significant relativistic effects in the third row of the periodic table. The valence *s* and *p* orbital contraction for Kr should raise some concern for the ability of a non-relativistic Hamiltonian to properly describe the electronic structure for elements in or beyond the third row, while the extreme difference in the case of Rn gives strong evidence that completely non-relativistic techniques are not appropriate for systems containing elements in this region of the periodic table.

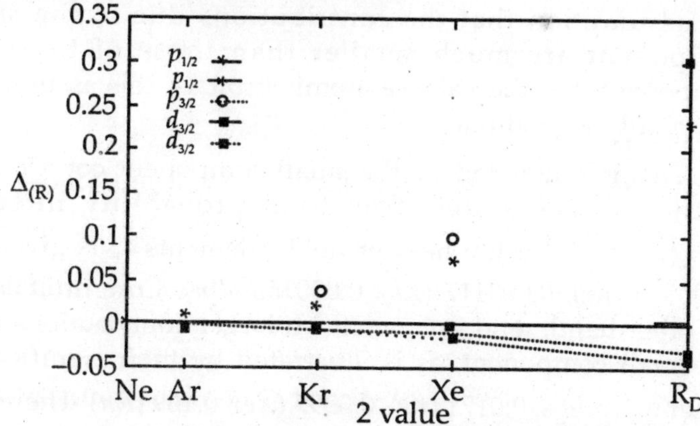

Fig. Difference between the $\langle R_{nrel} \rangle$ obtained with the non-relativistic atomic orbitals and $\langle R_{rel} \rangle$ obtained with the relativistic valence orbitals for the noble gas elements.

The horizontal trends of the third, fourth and fifth rows, illustrated in Figures support the "gold maximum" which is often observed for relativistic effects. Those elements which are closer to the center of the periodic table clearly exhibit stronger *s* and *p* orbital contraction and *d* orbital expansion than do those which are closer to the noble elements.

The $p_{1/2}$ orbitals display a greater orbital contraction than does the $S_{1/2}$ for some elements. The overall contraction of the *s* shell is still greater than that for the *p* shell, since the $p_{3/2}$ exhibit only a minute contraction, and, in some cases, even a slight expansion.

Another test of the necessity of employing a relativistic description of the electronic structure is the magnitude of the contribution of the small component of the wavefunction. Many approximate, quasi-relativistic methods employ only the large component of the wavefunction in the definition of some contributions to the Hamiltonian, as is done in many implementations of relativistic effective core potentials. Other methods, such as Douglas-Kroll based methods, perform a decoupling of the large and small component, and rely upon the decoupled large component to describe the electronic wavefunction. The success of these methods is dependent upon the assumption that the contributions due to the small component are much smaller than those of the large component. For the valence atomic orbitals, this assumption is typically a good one.

This is illustrated by the small component contribution to the valence **s** s-electron total probability integral: $\eta_s = \langle \phi_s | \phi_s \rangle$. For the heavier noble elements η_s is given by: 0.000116 (*Kr*), 0.000117 (*Xe*), 0.000153 (*Rn*). Core orbitals, on the other hand, tend to have much larger contributions from the small component, as is illustrated by their significantly larger η_s values: 0.0171 (*Kr*), 0.0397 (*Xe*), 0.109 (*Rn*). Therefore, if accurate descriptions of the core orbitals are desired, fully relativistic 4-component methods should be applied to systems containing fourth and fifth row atoms.

Finally, spin-orbit effects in atomic systems can provide an especially important gauge of the necessity for a relativistic treatment. The radial expectation value differences reported in Tables clearly illustrate the magnitude of the orbital splitting between the valence $p_{1/2}$ and $p_{3/2}$ orbitals as well as the $d_{3/2}$ and $d_{5/2}$ orbitals. Clearly this splitting will have significant implications on the chemical behaviour predicted for heavy elements by the DHF model as compared to the HF model. The magnitude of the splitting follows the same general trends as is observed for the the orbital contraction of valence *s* and *p* orbitals. The effect is highly pronounced for the fourth and fifth rows with the elements closest to the transition metals exhibiting the strongest splitting. A similar trend is observed for the DHF orbital energies as compared to the HF orbital energies.

CHEMICAL CONSEQUENCES

Although relativity can play an important role in the prediction of atomic or molecular spectra for all elements, the chemical consequences of relativity typically become evident only in the fifth row, and occasionally in the fourth row. In the fifth row, the relativistic effects are so large as to make even quantitative assessment of the atomic or molecular electronic structure and chemical behaviour impossible without considering the orbital contraction, orbital expansion, and spin-orbit effects.

The major differences in chemical behaviour predicted by the relativistic model in comparison to the non-relativistic model are bond contraction and changes in valency. The bond contraction has been found to be directly related to the *s* and *p* orbital contraction, in so far as both are manifestations of the relativistic lowering of the kinetic energy in the core and intermolecular region. The changes in valency are rooted in the stabilization of *s* and *p* orbitals, the destabilization of *d* and *f* orbitals, and spin orbit splitting.

In the first and second columns, where the valence occupancy is not dependent upon the relative energetics of the outer shells, the primary relativistic effect is bond contraction. The relativistic contraction of the outer *s* and *p* shells of Fr and Ra is so great, that these species actually exhibit smaller atomic radii than the lighter Cs and Ba, respectively. The actual extent to which σ-bonds involving the valence *s* orbital of these elements are contracted relative to the non-relativistic case has not been established.

The third column of the main group elements exhibits some of the most interesting relativistic effects of all the elements in the main group. The valence occupancy of these elements is $(ns)^2(np)^1$. For all the elements up to and including In, the (*ns*) shell is relatively close in energy and spatial extent to the *np* shell, and these shells can hybridize to allow for trivalent bonding. For Tl, however, two effects are manifest: the *ns* shell is dramatically stabilized, relative to the more modest stabilization of the $np_{1/2}$ shell, while the $np_{1/2}$ and $np_{3/2}$ shells experience a strong spin-orbit splitting. The result is that the **s** electrons become relatively inert, and that Tl appears to be mono-valent.

Main group columns four through seven exhibit progressively less pronounced relativistic effects. Lead exhibits a preference for a valency of 2, in contrast to the lighter elements which exhibit a valency of 4. This change in valency can be explained in much the same fashion as was the valency of 1 for Tl, except that for lead, the spin-orbit splitting of the two *p* shells is not as prominent. Excellent explorations of the importance of relativistic effects in the chemistry of the group IV di- and tetra-hydrides have been performed by Dyall and others.

Heavy transition metals exhibit the strongest relativistic effects of all the elements. One of the most dramatic examples is that of the contrast between silver and gold. When the non-relativistic predictions of the electronic spectrum of silver and gold atoms are compared, no striking differences appear, and one would expect Au to behave in an extremely similar manner to Ag, both in its electronic spectrum and in its chemical interaction with other species. What is observed, however, is that gold absorbs strongly in the visible region, thereby giving the metal its yellow luster. An explanation of this difference can be found in the relativistic prediction of the electronic structure. Gold has a ground state occupation of $(5d)^{10}(6s)^1$. The relativistic destabilization of the 5*d* orbitals and the stabilization of the 6*s* narrows the energetic gap between the two orbitals, and allows for excitation of the *d* electrons within the visible range in the bulk metal. Another interesting manifestation of the large relativistic effects experience by the gold atom is found in the bond length of AuH. The Hartree-Fock method predicts an AuH bond length which is significantly longer than the AgH bond length in contrast to the experimental observation that AuH is shorter. The Dirac-Hartree-Fock method, on the other hand, predicts an AuH bond that is 0.2Å shorter than the predicted AgH bond, in qualitative agreement with experiment. This dramatic evidence of the necessity for relativistic treatment in heavy element chemistry has given coinage metal containing compounds a prominent role in relativistic quantum chemistry.

Chapter 7

The Born-Oppenheimer

We may write the nonrelativistic Hamiltonian for a molecule as a sum of five terms:

$$\widehat{H} = -\frac{\hbar^2}{2m}\sum_i \nabla_i^2 - \sum_A \frac{\hbar^2}{2M_A}\nabla_A^2 - \sum_{A,i}\frac{Z_A c^2}{r_{Ai}} +$$

$$\sum_{A>B}\frac{Z_A Z_B c^2}{R_{AB}} + \sum_{i>j}\frac{e^2}{r_{ij}},$$

where i, j refer to electrons and A, B refer to nuclei. In atomic units, this is just

$$\widehat{H} = -\frac{1}{2}\sum_i \nabla_i^2 - \sum_A \frac{1}{2M_A}\nabla_A^2 - \sum_{A,i}\frac{Z_A}{r_{Ai}} +$$

$$\sum_{A>B}\frac{Z_A Z_B}{R_{AB}} + \sum_{i>j}\frac{1}{r_{ij}}.$$

The Schrödinger equation may be written more compactly as

$$\widehat{H} = \widehat{T}_N(R) + \widehat{T}_e(r) + \widehat{V}_{eN}(r,R) + \widehat{V}_{NN}(R) + \widehat{V}_{ee}(r),$$

where R is the set of nuclear coordinates and r is the set of electronic coordinates. If spin-orbit effects are important, they can be added through a spin-orbit operator $\widehat{H}_{10}$.

Unfortunately, the $\hat{V}_eN(r,R)$ term prevents us from separating $\hat{H}$ into nuclear and electronic parts, which would allow us to write the molecular wavefunction as a product of nuclear and electronic terms, $\Psi(r,R) = \Psi(r) \times (R)$. We thus introduce the Born-Oppenheimer approximation, by which we conclude that this nuclear and electronic separation is approximately correct.

The term $\hat{V}_{eN}(r,R)$ is large and cannot be neglected; however, we can make the R dependence parametric, so that the total wavefunction is given as $\Psi(r;R) \times (R)$. The Born-Oppenheimer approximation rests on the fact that the nuclei are much more massive than the electrons, which allows us to say that the nuclei are nearly fixed with respect to electron motion. We can fix R, the nuclear configuration, at some value R_a, and solve for the electronic wavefunction $\Psi(r;R_a)$, which depends only parametrically on R. If we do this for a range of R, we obtain the potential energy curve along which the nuclei move.

We now show the mathematical details. Initially, $\hat{T}_N(R)$ can be neglected since is smaller than $\hat{T}_e$ by a factor of M_A / μ_e, where μ_e is the reduced mass of an electron. Thus for a fixed nuclear configuration, we have

$$\hat{H}_{el} = \hat{T}_e(r) + \hat{V}_{eN}(r;R) + \hat{V}_{NN}(R) + \hat{V}_{ee}(r)$$

such that

$$\hat{H}_{el}\Psi(r;R) = E_{el}(r;R)$$

This is the "clamped-nuclei" Schrödinger equation. Quite frequently $\hat{V}_{NN}(R)$ is neglected in the above equation, which is justified since in this case R is just a parameter so that $\hat{V}_{NN}(R)$ is just a constant and shifts the eigenvalues only by some constant amount. Leaving $\hat{V}_{NN}(R)$ out of the electronic Schrödinger equation leads to a similar equation,

$$\hat{H}_e = \hat{T}_e(r) + \hat{V}_{eN}(r;R) + \hat{V}_{ee}(r)$$

$$\hat{H}_e \Psi(r;R) = E_e \Psi_e(r;R)$$

For the purposes of these notes, we will assume that $\hat{V}_{NN}(R)$ is included in the electronic Hamiltonian. Additionally, if spin-orbit effects are important, then these can be included at each nuclear configuration according to

$$\hat{H}_0 = \hat{H}_{eI} + \hat{H}_{I0}$$

$$\hat{H}_0 \Psi(r;R) = E_0 \Psi_e(r;R)$$

Consider again the original Hamiltonian. An exact solution can be obtained by using an (infinite) expansion of the form

$$\Psi(r;R) = \sum_k \Psi_k(r;R) X_k(R),$$

although, to the extent that the Born-Oppenheimer approximation is valid, very accurate solutions can be obtained using only one or a few terms. Alternatively, the total wavefunction can be expanded in terms of the electronic wavefunctions and a set of pre-selected nuclear wavefunctions; this requires the introduction of expansion coefficients:

$$\Psi^i(r,R) = \sum_{kI} C^i_{kI} \Psi_k(r;R) X_{kI}(R)$$

where the superscript i has been added as a reminder that there are multiple solutions to the Schrödinger equation.

Expressions for the nuclear wavefunctions $X_k(R)$ can be obtained by inserting the expansion into the total Schrödinger equation yields

$$\hat{H} \sum_{k'} \Psi_{k'}(r;R) X_{k'}(R) = E \sum_{k''} \Psi_{k''}(r;R) X_{k''}(R)$$

or

$$\int dr \Psi^e_k(r;R) \hat{H} \sum_{k'} \Psi_{k'}(r;R) X_{k'}(R) = E X_k(R)$$

if the electron functions are orthonormal. Simplifying further,

$$EX_k(R) = \sum_{k'} \int dr \Psi_k^e(r;R)[\hat{T}_N + \hat{H}_{el} + \hat{H}_{I0}]\Psi_{k'}(r;R)X_{k'}(R)$$

$$= \sum_{k'} \left\{ \left\langle \Psi_k(r;R) | \hat{H}_{el} | \Psi_{k'}(r;R) \right\rangle + \left\langle \Psi_k(r;R) | \hat{H}_{I0} | \Psi_{k'}(r;R) \right\rangle \right\}$$

$$+ \sum_{k'} \int dr \Psi_k^e(r;R) \hat{T}_N \Psi_{k'}(r;R) X_{k'}(R)$$

The last term can be expanded using the chain rule to yield

$$\sum_A \frac{-1}{2M_A} \Big[\nabla_A^2 X_k(R) + 2\langle \Psi_k(r;R) \nabla_A | \Psi_k(r;R) \rangle \nabla_A X_k(R)$$

$$X_k(R) \left\langle \Psi_k(r;R) \nabla_A^2 | \Psi_k(r;R) \right\rangle \Big]$$

$$- \sum_A \frac{1}{2M_A} \sum_{k' \neq k} \big[2\langle \Psi_k(r;R) \nabla_A | \Psi_k(r;R) \rangle \nabla_A x_{k'}(R)$$

$$+ X_{k'}(R) \left\langle \Psi_k(r;R) \nabla_A^2 | \Psi_{k'}(r;R) \right\rangle \big].$$

At this point, a more compact notation is very helpful. Following Tully, we introduce the following quantities:

$$U_{kk'}(R) = U^{(el)}{}_{kk'}(R) + U^{(so)}{}_{kk'}(R)$$

$$U_{kk'}^{(el)}(R) = \left\langle \Psi_k(r;R) | \hat{H}_{el} | \Psi_{k'}(r;R) \right\rangle$$

$$U_{kk'}^{(so)}(R) = \left\langle \Psi_k(r;R) | \hat{H}_{so} | \Psi_{k'}(r;R) \right\rangle$$

$$\hat{T}_{kk'}^{\eta}(R) = \sum_A \frac{-1}{M_A} d_{kk'}^{(A)}(R).\nabla_A$$

$$\hat{T}_{kk'}^{\eta}(R) = \sum_A \frac{-1}{2M_A} D_{kk'}^{(A)}(R)$$

$$d_{kk'}^{(A)}(R) = \langle \Psi_k(r;R) | \nabla_A | \Psi_{k'}(r;R) \rangle$$

$$D_{kk'}^{(A)}(R) = \left\langle \Psi_k(r;R) \mid \nabla_A^2 \mid \Psi_{k'}(r;R) \right\rangle$$

$$\left[\hat{T}_N + \sum_A \left(\frac{-1}{2M_A}\right)\left\{d_{kk'}^{(A)}\nabla_A + D_{kk}^{(A)}\right\} + U_{kk} - E\right] X_k(R)$$

$$= -\sum_{k \neq k'}\left[U_{kk'} + \sum_A \left(\frac{-1}{2M_A}\right)\left\{2d_{kk'}^{(A)}\nabla_A + D_{kk'}^{(A)}\right\}\right] X'_k(R),$$

or

$$= \left[\hat{T}_N + \hat{T}'_{kk} + T''_{kk} + U_{kk} - E\right] X_k(R)$$

$$= -\sum_{k' \neq k}\left[U_{kk'} + \hat{T}'_{kk'} + T''_{kk'}\right] X_{k'}(R)$$

This is equation of Tully's article. Tully simplifies this equation by one more step, removing the term $\hat{T}'_{kk}$. By taking the derivative of the overlap of $\left\langle \Psi_k(r;R) \mid \Psi_k(r;R) \right\rangle$ it is easy to show that this term must be zero when the electronic wavefunction can be made real. If we use electronic wavefunctions which diagonalize the electronic Hamiltonian, then the electronic basis is called adiabatic, and the coupling terms $U_{kk'}$ vanish.[1] This is the general procedure. However, the equation above is formally exact even if other electronic functions are used. In some contexts it is preferable to minimize other coupling terms, such as $\hat{T}'_{kk'}$; this results in a diabatic electronic basis. Note that the first-derivative nonadiabatic coupling matrix elements $\hat{T}'_{kk'}$ are usually considered more important than the second-derivative ones, $T''_{kk'}$.

In most cases, the couplings on the right-hand side of the preceeding equation are small. If they can be safely neglected, and assuming that the wavefunction is real, we obtain the following equation for the motion of the nuclei on a given Born-Oppenheimer potential energy surface:

$$\left[\hat{T}_N + T''_{kk} + U_{kk}\right] X_k(R) = E_{X_k}(R)$$

This equation clearly shows that, when the off-diagonal couplings can be ignored, the nuclei move in a potential field set up by the electrons. The potential energy at each point is given primarily by U_{kk} (the expectation value of the electronic energy), with a small correction factor T''_{kk}. Following Steinfeld, we can estimate the magnitude of the term T''_{kk} as follows: a typical contribution has the form $1/(2M_A)\nabla_A^2\Psi(r;R)$, but $\nabla_A\Psi(r;R)$ is of the same order as $\nabla_i\Psi(r;R)$ since the derivatives operate over approximately the same dimensions.

The latter is $\Psi(r;R)p_e$, with p_e the momentum of an electron. Therefore $1/(2M_A)\nabla_A^2\Psi(r;R) \approx p_e^2/(2M_A) = (m/M_A)E_e$. Since $m/M_A \sim 1/10000$, this term is expected to be small, and it is usually dropped. For methylene, this term, called the Born-Oppenheimer diagonal correction (BODC), has an effect of about 40 cm^{-1} to the singlet-triplet energy gap.

THE HYDROGEN ATOM

In 1885 a Swiss secondary school teacher named Johann Jacob Balmer published a short note (entitled "Note on the Spectral Lines of Hydrogen", *Annalen der Physik und Chemie* 25, 80-5) in which he described an empirical formula for the four most prominent wavelengths of light emitted by hydrogen gas. These wavelengths had been measured with great precision by Vogel and Huggins, giving the four values 6562.10, 4860.74, 4340.10, and 4101.20 Angstroms (10^{-10} m). Balmer's note does not make clear whether he was also aware of the measured series limit, λ_∞ = 3645.6 A, or whether he deduced this himself. In any case, one can find by numerical experimentation that the four characteristic wavelengths are closely proportional to the following products of small primes

$$1512 = 2^3.3^3.7,\ 1120 = 2^5.5.7,\ 1000 = 2^3.5.7,\ 945 = 3^3.5.7$$

Three of these are divisible by 2^3, three are divisible by 5, three are divisible by 7, and two are divisible by 3^3. Thus we can easily express these numbers as simple fractional multiples of $840 = 2^3 \times 3 \times 5 \times 7$, which corresponds to the series limit $\lambda_\infty = 3645.6$ A. It may have been just this kind of numerical experimentation that led Balmer to recognize that the four prominent wavelengths are given very closely by $(9/5)\lambda_\infty$, $(16/12)\lambda_\infty$, $(25/21)\lambda_\infty$, and $(36/32)\lambda_\infty$. He also noticed that the numerators of the coefficients are consecutive squares, and each denominator is 4 less than the numerator. He speculated that the pattern would continue up to the series limit, which is indeed the case. In terms of the wave number k $(=1/\lambda)$, Balmer's formula can be written as

$$k = \frac{n^2}{\lambda_\infty}\left(\frac{j^2 - n^2}{n^2 j^2}\right) = R_H\left(\frac{1}{n^2} - \frac{1}{j^2}\right) \quad n = 2; j = 3, 4, 5, \ldots$$

where $R = n^2/\lambda_\infty$ with $n = 2$. The parameter R_H is now called Rydberg's constant for hydrogen, and the best empirical value is 10967757.6 m^{-1}. As Balmer also speculated, if we take different values of n we get different series of spectral lines. The series with n = 1, 2, 3, 4, and 5 are now known as the Lyman, Balmer, Paschen, Brackett, and Pfund series, respectively, which characteristic the spectral lines of the hydrogen atom. (The Balmer series was observed first because its frequencies are in the visible and near ultra-violet range.) This is an outstanding example of a successful empirical fit (like Bode's Law in astronomy) for a class of physical phenomena that was not based on any underlying physical model or theory, i.e., no reason was known for why wavelengths of light emitted from a hydrogen atom should exhibit this pattern.

In classical terms a hydrogen atom consists of a proton and an electron bound together by their mutual electrical attraction. To keep them from collapsing together, we might imagine that the electron is revolving in "orbit" around the proton, similar to a planet revolving around the Sun, with the centrifugal force balancing the electrical attraction. However, this model is not satisfactory, because the electron would be continuously accelerating, and according to classical theory an accelerating charge radiates energy in the form of electromagnetic waves. As a result, the orbiting electron would very quickly radiate away all of its kinetic energy and spiral into the proton. Thus the existence of stable atoms was inexplicable in the context of classical physics, as was the characteristic set of discrete energy levels of atoms.

By the early 1900's it had become clear that classical electrodynamics was inadequate to account for the behaviour of either the electromagnetic field or of elementary particles. In 1900 Max Planck had shown in his study of black-body radiation that it is necessary to quantize the energy of electromagnetism in order to avoid the "untra-violet catastrophe", and he introduced the fundamental constant h. In 1905 Einstein made the even more radical proposal that in some respects electromagnetic wave energy propagates as if it consists of small packets (photons) with many of the characteristics of particles, each photon having an energy E related to the wave frequency v by $E = hv$.

In 1913 Niels Bohr developed a new representation of the hydrogen atom by combining classical ideas with a few additional postulates that were suggested by the nascent quantum concepts of Planck and Einstein. First, he assumed that the angular momentum of an electron in orbit around the nucleus must be an integer multiple of $\hbar$ (Planck's constant h divided by 2π). It follows that only a certain set of discrete energy levels may occur. Second, he assumed that an electron radiates energy only when it makes a transition from one stable orbit to another of lower energy. If ΔE is the difference in energy levels, then he assumed that the transition resulted in the emission of a photon with this amount of energy, and hence with the frequency $v = \Delta E/h$ in accord with Einstein's postulate. Armed with these postulates, Bohr reasoned that an electron of mass m orbiting a proton (of much greater mass) at radius r would satisfy the classical force balance.

$$\frac{e^2}{4\pi \epsilon_0 r^2} = m\frac{v^2}{r}$$

and the total energy (kinetic plus potential) has the classical value

$$E = \frac{1}{2}mv^2 + (-e)V$$

$$= \frac{e^2}{8\pi \epsilon_0 r} - \frac{e^2}{4\pi \epsilon_0 r} = -\frac{e^2}{8\pi \epsilon_0 r}$$

Likewise the angular momentum has the classical value

$$L = mvr = \sqrt{\frac{me^2 r}{4\pi \epsilon_0}}$$

Bohr then imposed his quantization assumption, asserting that L must be an integer multiple of $\hbar$. Setting $L = n\hbar$ and solving the above equation for r, we get

$$r = \frac{4\pi \epsilon_0 n^2 \hbar^2}{me^2}$$

Substituting into the expression for the energy E gives the corresponding quantized energy levels of the hydrogen atom

$$E_n = -\frac{me^4}{2(4\pi \epsilon_0)\hbar^2 n^2}$$

This provides a nice rationale for Balmer's empirical formula, because it implies that the frequency of the emitted light when an electron makes a transition from the jth to the nth energy level is

$$\nu = \frac{\Delta E}{h} = \frac{me^4}{2h(4\pi \epsilon_0)^2 \hbar^2}\left(\frac{1}{n^2} - \frac{1}{j^2}\right)$$

Since $\lambda\nu = c$ we have $K = \nu/c$ and therefore

$$K = \left(\frac{me^4}{h^2 \varepsilon_0{}^2 c}\right)\left(\frac{1}{n^2} - \frac{1}{j^2}\right)$$

Actually, to be more accurate, the mass m in this expression should be replaced with the "reduced mass" $mM/(m + M)$ where M is the mass of the proton (or, more generally, the nucleus), just as in classical orbital mechanics. Then the coefficient in the above expression is identified with Rydberg's constant R_H for the hydrogen atom, and using the values of the fundamental constants

$$c = 2.99792458.10^{8} \text{ m/s}$$

$$h = 6.626176.10^{-34} \text{ J.s}$$

$$\varepsilon_0 = 8.85418788.10^{-12}\ F/m$$

$$e = 1.6021892.10^{-19} \text{ C}$$

$$m = 9.109534.10^{-31} \text{ kg}$$

$$M = 1.6726485.10^{-27} \text{ kg}$$

we can compute

$$R_H = \left(\frac{mM}{m+M}\right)\frac{e^4}{h^3 \varepsilon_0{}^2 c} = 10967757.78 \text{ m}^{-1}$$

This is in remarkable agreement with the measured value of $10967757.6 \pm 1.2 \text{ m}^{-1}$ from spectroscopic data. Nevertheless, Bohr's model of the atom is not completely satisfactory, partly because of the ad hoc nature of its premises, and also because it's representation of electrons as tiny particles with definite trajectories is not viable in a wider context.

A somewhat plausible justification for Bohr's quantization postulate came in 1924 when Louis de Broglie developed the idea that particles of matter on the smallest scale exhibit wave-like properties, complementing Einstein's suggestion that electro-magnetic waves exhibit particle-like properties. The de Broglie wavelength for the matter wave corresponding to a particle with momentum p is $\lambda = h/p$, and if we stipulate that the circumference $2\pi r$ of a circular orbit of radius r must be an integer multiple of the wavelength, we have $2\pi r/l = 2\pi r\pi/h = n$ for some positive integer n. Since the angular momentum is $L = pr$, this immediately gives Bohr's quantization postulate $L = n\hbar$. This "orbital wave quantization" is illustrated for $n = 1$ through 5 in the figures below.

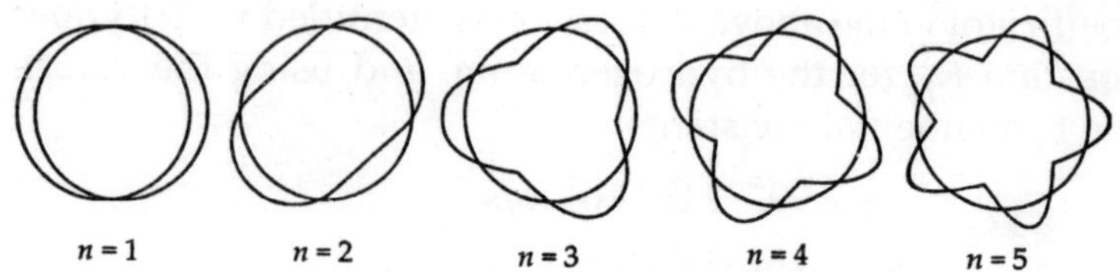

However, despite the plausibility of this approach, Bohr's model of the hydrogen atom, even with de Broglie's justification and with subsequent refinements by Sommerfeld, is now considered obsolete, having been superceded by a more thorough-going wave mechanics developed by Erwin Schrodinger in 1925. (This new theory was subsequently shown to be essentially identical to the "matrix mechanics" already developed by Werner Heisenberg in 1924.)

Schrodinger's wave mechanics postulates that a particle is characterized by a complex-valued wave function $\Psi(x, y, z, t)$ whose squared norm at any point equals the probability density for the particle to be found at that point. (The probability interpretation of Schrodinger's wave function was first proposed by Max Born.) In addition, Schrodinger postulated that, in a region where there is a potential field $V(x, y, z, t)$, the wave function Ψ of a particle is governed by the equation

$$-\frac{\hbar^2}{2m}\left[\frac{\partial^2\Psi}{\partial x^2}+\frac{\partial^2\Psi}{\partial y^2}+\frac{\partial^2\Psi}{\partial z^2}\right]+V\Psi=i\hbar\frac{\partial\Psi}{\partial t}$$

It's possible to give a plausibility argument for this equation, but here we will just take it as given. Expressing the spatial Laplacian (the quantity in the square brackets) in terms of polar coordinates, and considering just the radial part, this equation is

$$-\frac{\hbar^2}{2m}\left[\frac{1}{r^2}\frac{\partial}{\partial r}\left(r^2\frac{\partial\Psi(r,t)}{\partial r}\right)\right]+V(r,t)\Psi(r,t)=i\hbar\frac{\partial\Psi(r,t)}{\partial t}$$

Furthermore, if the potential field V does not change with time, we can separate the variables by expressing $\Psi(r, t)$ as a product of a spatial part and a temporal part, i.e., we have functions $\varphi(r)$ and $\omega(t)$ such that $\Psi(r, t) = \varphi(r)\omega(t)$. Substituting into the above equation and dividing through by $\Psi(r, t)$ gives

$$-\frac{\hbar^2}{2m\varphi(r)}\left[\frac{1}{r^2}\frac{d}{dr}\left(r^2\frac{d\varphi(r)}{dr}\right)\right]+V(r)=\frac{i\hbar}{\omega(t)}\frac{d\omega(t)}{dt}$$

Since r and t are independent, the left and right hand sides of this equation must both equal a constant, which we will call E, so the time-independent radial Schrodinger equation in this simple case is

$$-\frac{\hbar^2}{2m}\left[\frac{1}{r^2}\frac{d}{dr}\left(r^2\frac{d\varphi(r)}{dr}\right)\right]+V(r)\varphi(r)=E\varphi(r)$$

Now, as mentioned previously, for the region around a charged proton the potential energy due to the Coulomb force is given by

$$V(r)=\frac{e^2}{4\pi\varepsilon_0 r}$$

Inserting this into the Schrodinger equation, evaluating the nested derivatives, and re-arranging terms, we get

$$\frac{d^2\varphi}{dr^2}+\frac{2}{r}\frac{d\varphi}{dr}+\frac{2m}{\hbar^2}\left[\frac{e^2}{4\pi\varepsilon_0 r}+E\right]\varphi=0$$

For sufficiently large values of r the terms with r in the denominator will be negligible, so the equation will reduce to

$$\frac{d^2\varphi}{dr^2}+\frac{2m}{\hbar^2}E\varphi=0$$

which has the solution

$$\varphi(r)=e^{-Cr} \text{ where } C=\sqrt{-\frac{2mE}{\hbar^2}}$$

To exploit this asymptotic result, and without loss of generality, we can consider a general solution of the form $\varphi(r) = F(r)e^{-Cr}$. Substituting into the Schrodinger equation, evaluating the derivatives, and dividing through by e^{-Cr}, we get

$$\frac{d^2F}{dr^2}+2\left(\frac{1}{r}-C\right)\frac{dF}{dr}+\frac{2}{r}(B-C)F=0$$

where

$$B=\frac{me^2}{4\pi\varepsilon_0\hbar^2}$$

If the function $F(r)$ is analytic it can be represented by a power series, i.e., it can be expressed in the form

$$F(r) = f_0 + f_1 r + F_2 r^2 + ...$$

Substituting into the equation for F, collecting terms by powers of r, and setting the coefficients of these terms to zero, we arrive at the conditions

$$f_k = 2\frac{kC-B}{k(k+1)}f_{k-1}$$

for $k = 1, 2, ...$ For sufficiently large k these expressions approach $f_k = [2C/(k + 1)]f_{k-1}$, which is the series for e^{2Cr}, and hence $F(r)e^{-Cr}$ goes to e^{Cr}, which increases to infinity as r increases. Therefore, in order to give a solution that goes to zero as r goes to infinity, we must impose the requirement that the series for $F(r)$ terminates after a finite number of terms. This occurs if and only if $nC = B$ for some positive integer n. Hence the necessary and sufficient condition for the solution to approach zero as r increases is

$$n\sqrt{-\frac{2mE}{\hbar^2}}=\frac{me^2}{4\pi\varepsilon_0\hbar^2}$$

Squaring both sides and solving for E gives the allowable energy levels

$$E_n=-\frac{me^4}{(4\pi\varepsilon_0)^2 2\hbar^2 n^2}$$

which is identical to the discrete energy levels of Bohr's model discussed previously. It's worth noting that the quantization of energy levels here is not the result of quantized angular momentum or orbital standing-waves. It arises from an analysis of the purely radial component of the Schrodinger wave equation of the ground state, which is spherically symmetrical and has an angular quantum number of zero.

Superficially it isn't obvious that the "realistic solutions" of (1) must be quantized, so it's worthwhile to examine the solution technique more closely to understand clearly how the quantization arises. First, notice that if we had tried to find a series solution of (1) directly by inserting a series $\varphi(r) = \varphi_0 + \varphi_1 r + \varphi_2 r^2 + \ldots$ we would have arrived at a set of conditions involving three consecutive coefficients. This can be seen by inspection, because when we carry out the differentiations and collect the coefficients of r^k, any term of the original differential equation of the form $r^s\, d^q\varphi/dr^q$ contributes a quantity involving φ_{k+q-s}.

Hence the four terms of (1) contribute quantities involving φ_{k+2}, φ_{k+2}, φ_{k+1}, and φ_k respectively. In contrast, the four terms of (2) contribute quantities involving f_{k+2}, f_{k+2}, f_{k+1}, and f_{k+1} respectively, so the power series conditions enable us to determine each f_{k+2} as a multiple of f_{k+1}. We originally motivated the solution form $F(r)e^{-cr}$ based on the asymptotic solution for large r, but we could also have justified it based on the fact that this transformation leads to a differential equation whose power series solution is subjected to conditions on just two consecutive coefficients. The fact that such a transformation exists is crucial for the quantization.

This leads us to consider the general conditions in which such a transformation exists. Suppose we have a second-order differential equation of the form

$$\alpha(x)\frac{d^2y}{dx^2} + \beta(x)\frac{dy}{dx} + \gamma(x)y = 0$$

where α, β, and γ are rational functions of x. Since we can multiply through by arbitrary polynomials in x, we can assume without loss of generality that α, β, γ are polynomials in x. If we postulate a solution of the form $y(x) = f(x)e^{g(x)}$ and substitute into this equation, we get

$$[\alpha]\ddot{f}+[2\alpha\dot{g}+\beta]\dot{f}+[\alpha\ddot{g}+\alpha\dot{g}^2+\beta\dot{g}+\gamma]f=0$$

In order for the series solution for $f(x)$ to have conditions on just the sets of two consecutive coefficients, there must be an integer d such that the coefficient of f'' contains only terms in x^{d+1} and x^{d+2}, and the coefficient of f' contains only terms in x^d and x^{d+1}, and the coefficient of f contains only terms in x^d. We seek a function $g(x)$ such that these conditions are satisfied. In the case of equation (1) we have (after multiplying through by r) an equation of the form (5) with $a(x) = x$, $b(x) = 2$, and $g(x) = 2B - C^2x$ where B and C are the constants defined previously.

Therefore, from the condition on a, we see that d is either 0 or 1, so we need a function $g(x)$ such that $2xg' + 2$ involves only terms in x^0 and x^1, or only terms in x^1 and x^2. Since it certainly involves a term in x^0, the remaining term must be in x^1, so $g'(x)$ must be a constant. Also, the coefficient of f must be in x^0 (i.e., a constant), so we have $x(g')^2 + 2g' + 2B - C^2x = K$. Therefore we must have $(g')^2 = (C)^2$, which leads to the transformation $y(x) = f(x)e^{-Cx}$ as expected.

The superiority of the wave mechanical model of the hydrogen atom over Bohr's model is immense, because it not only duplicates and (in a sense) "explains" the quantized energy levels, it actually gives the complete probability density functions for the various possible stationary states. Using the recursive formula (3) we can evaluate the coefficients of the polynomial $F(r)$ for each value of the quantum number n. Combining these polynomials with the exponential parts, we have the wave functions of the first few states

$$\varphi_1(r) = f_0 e^{-Cr}$$

$$\varphi_2(r) = (1 - Cr) f_0 e^{-Cr}$$

$$\varphi_3(r) = \left(1 - 2Cr + \frac{2}{3}C^2r^2\right) f_0 e^{-Cr}$$

$$\varphi_4(r) = \left(1 - 3Cr + 2C^2r^2 - \frac{1}{3}C^3r^3\right) f_0 e^{-Cr}$$

$$\varphi_5(r)=\left(1-4Cr+4C^2r^2-\frac{4}{3}C^3r^3+\frac{2}{15}C^4r^4\right)f_0e^{-Cr}$$

Recall that $B = nC$ and B is composed entirely of fundamental constants, independent of n, and it has units of inverse length. Letting a_0 denote the length $1/B = 1/(nC)$, and using m instead of the more accurate reduced mass, we have

$$a_0=\frac{4\pi\varepsilon_0\hbar^2}{me^2}$$

and we can substitute $1/(na_0)$ for C in the preceding wave functions and clear fractions. In terms of the normalized radius parameter $\rho = r/a_0$ the wave functions are

$$\varphi_1(\rho)=f_0e^{-\rho}$$

$$\varphi_2(\rho)=(2-\rho)\frac{f_0}{2}e^{-\rho/2}$$

$$\varphi_3(\rho)=(27-18\rho+2\rho^2)\frac{f_0}{27}e^{-\rho/3}$$

$$\varphi_4(\rho)=(192-144\rho+24\rho^2-\rho^3)\frac{f_0}{192}e^{-\rho/4}$$

$$\varphi_5(\rho)=(9375-7500\rho+1500\rho^2-100\rho^3+2\rho^4)\frac{f_0}{9375}e^{-\rho/5}$$

Each of these wave functions includes a constant factor f_0. To determine the value of this factor, recall that the squared norm of the wave function is the probability density, and so the integral of this quantity over all of space must equal 1. The volume of an incremental spherical shell of radius r and thickness dr is $4pr^2dr$ so the probability integral is

$$\int_0^\infty 4\pi r^2(\varphi_n(r))^2\,dr=4\pi a_0{}^3\int_0^\infty \rho^2(\varphi_n(\rho))^2d\rho=1$$

For example, to find the constant factor f_0 for the case n = 1 we insert the wave function into this equation and evaluate the integral to give

$$\pi a_0^3 f_0^2 \int_0^\infty \rho^2 (e^{-\rho})^2 d\rho = \pi a_0^3 f_0^2 = 1$$

Therefore we have $f_0 = (\pi a_0^3)^{-1/2}$, and the complete wave function for the ground state of the hydrogen atom is

$$\varphi_1(r) = \frac{e^{-r/a_0}}{\sqrt{\pi a_0^3}}$$

Of course, in accord with our choice of nomalizing factors, the probability density for finding the electron in an incremental shell of radius r is $\delta_n(r) = 4\pi r^2 \varphi_n(r)^2$. This is plotted in the figure below for the first few values of n.

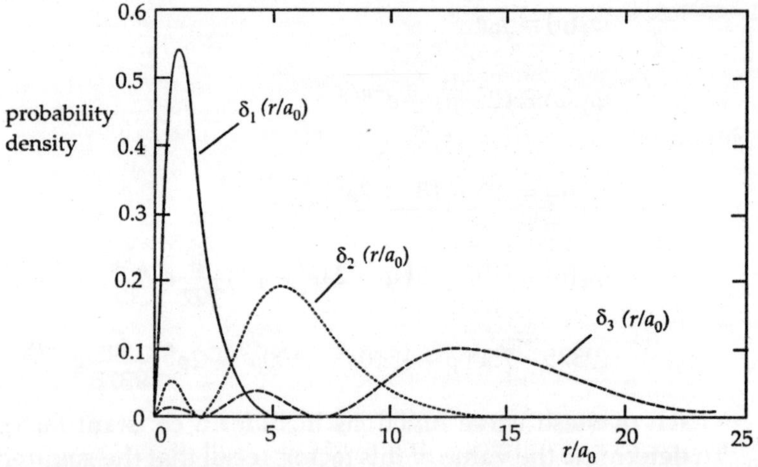

The angular components of the electron's wave function, effectively assuming that it has zero angular momentum, so the only non-zero quantum number was the radial one. This was all based on taking just the radial part of the Laplacian in the Schrodinger equation. If we had taken the angular parts we would have found that those too are associated with quantum numbers 0, 1, 2, ..., and they contribute to the overall orbital wave function. However, the essential features of Schrodinger's approach to the hydrogen atom are already apparent in the purely radial part.

Hydrogen is the simplest atom (one proton and one electron), but is still an extremely rich topic for study. The wave-function aspect of quantum mechanics prevents the electron from collapsing into an infinitesimally small orbit with an infinite negative energy. In classical physics, an attractive ~$1/r$ potential could have arbitrarily small orbitals with arbitrarily large binding energies. In quantum physics, the lowest energy orbital is called the ground state. Solving the Schroedinger equation for an electron in an attractive Coulomb potential provides the ground state properties of the hydrogen atom.

Table: Ground state properties of hydrogen

	Average orbital radius	**Binding energy**
ground state of	$r_0 = \frac{h^2}{4\pi^2 mke^2}$	$E_0 = \frac{-ke^2}{2r_0}$
hydrogen atom	$= 5.29.10^{-11}\ m$	$= -13.6\ eV$
ground state of higher	$r_0(Z) = \frac{5.29.10^{-11} m}{Z}$	$E_0(Z) = Z^2.$
charge atom		$(-13.6\ \text{eV})$
excited state of higher		$E_n(Z) = \frac{Z^2}{n^2}.$
charge atom		**$(-136\ eV)$**

Just to be confusing, the subscript 0 which refers to the ground state refers to $n = 1$, not $n = 0$.The radius of an electron in the ground state of hydrogen is called the Bohr radius, named after Niels Bohr, a famous Danish physicists. He found the radius by asking the question, "What circular orbit would have a momentum such that the circumference of the orbit would be exactly one DeBroglie wavelength?"

When electrons make a jump from one energy level (a specific value of n) to another energy level, a photon is emitted which carries away the necessary energy to satisfy energy conservation. The inverse wavelength (which is proportional to the energy) of the light is then propotional to the change in energy:

$$\frac{1}{\lambda}=R\left(\frac{1}{n^2}-\frac{1}{m^2}\right), R=1.097.10^7\ m^{-1}$$

R is known as Rydberg's constant. As an exercise, students can solve for R in terms of fundamental constants. The beginning level is denoted by m and the final level is denoted by n. For n = 1, the various wavelengths (m = 2, 3, 4 ...) are known as the Lyman series. The Balmer series is the series where n = 2 and the Paschen series is for n = 3. The values of the wavelengths are often called spectral lines because of the lines that appear in diffraction experiments. For a hot hydrogen source, where the atoms are excited thermally and then deexcite via the emission of photons, the Balmer series is in the visible range.

Spherical polar coordinates. The solution of the Schrodinger equation for the hydrogen atom is a formidable mathematical problem, but is of such fundamental importance that it will be treated in outline here. The solution is managed by separating the variables so that the wavefunction is represented by the product:

The separation leads to three equations for the three spatial variables, and their solutions give rise to three quantum numbers associated with the hydrogen energy levels.

HYDROGEN SCHRODINGER EQUATION

The electron in the hydrogen atom sees a spherically symmetric potential, so it is logical to use spherical polar coordinates to develop the Schrodinger equation. The potential energy is simply that of a point charge:

The expanded form of the Schrodinger equation is shown below. Solving it involves separating the variables into the form

$$\Psi(r,\theta,\phi)=R(r)P(\theta)F(\phi)$$

The starting point is the form of the Schrodinger equation:

$$\frac{-\hbar^2}{2\mu}\frac{1}{r^2\sin\theta}\left[\sin\theta\frac{\partial}{\partial r}\left(r^2\frac{\partial\Psi}{\partial r}\right)+\frac{\partial}{\partial\theta}\left(\sin\theta\frac{\partial\Psi}{\partial\theta}\right)+\frac{1}{\sin\theta}\frac{\partial^2\Psi}{\partial\phi^2}\right]$$

$$+U(r)\Psi(r,\theta,\phi)=E\Psi(r,\theta,\phi)$$

THE SCHRODINGER EQUATION IN 3-D: THE HYDROGEN ATOM

The Schrodinger equation in 1-D is extended to 3-D in much the same way as the energy method in 1-D is extended to 2-D to deal with central-force motion. The kinetic-energy term associated with the angular variable(s) is written as a function of r and the angular momentum l, and is treated as a form of potential energy. The resulting equation looks (somewhat in the case of the Schrodinger equation) like a 1-D problem with an "effective" potential energy.

The Bohr model of the hydrogen atom uses the classical techniques of central-force motion along with quantized angular momentum, and finds the energy levels successfully. It is a convenient starting point here because it provides the quantities we need to scale the Schrodinger equation into dimensionless form. The three equations needed for the model are Newton's law:

$$\frac{mV^2}{r}=\frac{e^2}{4\pi\varepsilon_0 r^2}$$

Energy conservation:

$$\frac{mV^2}{2}-\frac{e^2}{4\pi\varepsilon_0 r}=E$$

Quantization of the angular momentum:

$$l=mVr=n\hbar$$

where n has integral values 1, 2, 3.. The three equations can be solved for the radii and energies of the orbits:

$$r=n^2\frac{4\pi\varepsilon_0\hbar^2}{me^2}=n^2a_0$$

$$E = -\frac{me^4}{2(4\pi\varepsilon_0\hbar)^2}\frac{1}{n^2}$$

$$= -\frac{mc^2}{2}\left(\frac{e^2}{4\pi\varepsilon_0\hbar c}\right)^2\frac{1}{n^2} = -\frac{mc^2\alpha^2}{2}\frac{1}{n^2} = -\frac{13.6\ eV}{n^2}$$

The dimensionless group of constants represented by a is known as the fine-structure constant and is approximately equal to 1/137. We will use the first orbit radius (approximately 0.053 nm) and energy as the length and energy units in the dimensionless Schrodinger equation.

The Schrodinger equation looks fairly complicated when written in spherical polar coordinates. The solution is written as the product of a function R of r and a function Y (known as a spherical harmonic) of the angular variables. There are elegant ways of dealing with spherical harmonics in the quantum mechanical theory of angular momentum, and there are even more elegant ways in group theory where they are basis functions for the rotation group in 3-D. (There is nothing "quantum mechanical" about the spherical harmonics.) We will say nothing more about the angular variables here beyond noting that they contribute the angular momentum term to the radial Schrodinger equation:

$$-\frac{\hbar^2}{2m}\frac{d^2}{dr^2}(rR) + \frac{l(l+1)\hbar^2}{2mr^2}(rR) - \frac{e^2}{4\pi\varepsilon_0 r}(rR) = E(rR)$$

The angular momentum quantum number l has values 0, 1, 2.., and note the product $l(l + 1)$ in the angular momentum term rather than the n^2 of the Bohr model. It is not necessary to write the equation in terms of the product rR, but it is convenient. (The product has some physical significance: when it is squared it is proportional to the radial probability density, the probability density for a thin spherical shell at radius r.) In dimensionless form the equation becomes:

$$-\frac{d^2}{dr^2}(rR) + \frac{l(l+1)}{r^2}(rR) - \frac{2}{r}(rR) = E(rR)$$

This equation is not suitable for numerical solution because, even with one power of r cancelled, the angular momentum term goes to infinity at the origin. What this tells us is that near the origin,

$$R = r^l u$$

where u is non-zero at the origin. When we make this substitution,

$$\frac{d^2u}{dr^2} = -Eu - 2\frac{u + (l+1)du/dr}{r}$$

This is the equation we solve numerically using the Feynman algorithm. (The term in $1/r$ dictates the value of du/dr at the origin.) The programme operates in two modes:

1. It finds the energy levels for which the wavefunction tends asymptotically to zero at large r, and
2. It plots the radial probability density.

The usual *s, p, d, f* notation is used for l = 0, 1, 2, 3. The horizontal axis is divided into four divisions, each equal to the radius of the nth Bohr orbit. Suitable vertical scale factors are input along with the n and l values for the selected state.

The Motivation for Quantum Mechanics

Physicists at the end of the nineteenth century believed that most of the fundamental physical laws had been worked out. They expected only minor refinements to get "an extra decimal place" of accuracy. As it turns out, the field of physics was transformed profoundly in the early twentieth century by Einstein's discovery of relativity and by the development of quantum mechanics. While relativity has had fairly little impact on chemistry, all of theoretical chemistry is founded upon quantum mechanics.

The development of quantum mechanics was initially motivated by two observations which demonstrated the inadeqacy of classical physics. These are the "ultraviolet catastrophe" and the photoelectric effect.

THE ULTRAVIOLET CATASTROPHE

A *blackbody* is an idealized object which absorbs and emits all frequencies. Classical physics can be used to derive an equation which describes the intensity of blackbody radiation as a function of frequency for a fixed temperature—the result is known as the Rayleigh-Jeans law. Although the Rayleigh-Jeans law works for low frequencies, it diverges as v^2; this divergence for high frequencies is called the ultraviolet catastrophe.

Max Planck explained the blackbody radiation in 1900 by assuming that the energies of the oscillations of electrons which gave rise to the radiation must be proportional to integral multiples of the frequency, i.e.,

$$E = nhv$$

Using statistical mechanics, Planck derived an equation similar to the Rayleigh-Jeans equation, but with the adjustable parameter h. Planck found that for h = 6.626×10^{-34} J s, the experimental data could be reproduced. Nevertheless, Planck could not offer a good justification for his assumption of energy quantization. Physicicsts did not take this energy quantization idea seriously until Einstein invoked a similar assumption to explain the photoelectric effect.

THE PHOTOELECTRIC EFFECT

In 1886 and 1887, Heinrich Hertz discovered that ultraviolet light can cause electrons to be ejected from a metal surface. According to the classical wave theory of light, the intensity of the light determines the amplitude of the wave, and so a greater light intensity should cause the electrons on the metal to oscillate more violently and to be ejected with a greater kinetic energy. In contrast, the experiment showed that the kinetic energy of the ejected electrons depends on the *frequency* of the light. The light intensity affects only the number of ejected electrons and not their kinetic energies.

Einstein tackled the problem of the photoelectric effect in 1905. Instead of assuming that the electronic oscillators had energies given by Planck's formula, Einstein assumed that the radiation itself consisted of packets of energy $E = hv$, which are now called photons. Einstein successfully explained the photoelectric effect using this assumption, and he calculated a value of h close to that obtained by Planck.

Two years later, Einstein showed that not only is light quantized, but so are atomic vibrations. Classical physics predicts that the molar heat capacity at constant volume (C_v) of a crystal is $3R$, where R is the molar gas constant. This works well for high temperatures, but for low temperatures C_v actually falls to zero. Einstein was able to explain this result by assuming that the oscillations of atoms about their equilibrium positions are quantized according to $E = nh\nu$, Planck's quantization condition for electronic oscillators. This demonstrated that the energy quantization concept was important even for a system of atoms in a crystal, which should be well-modeled by a system of masses and springs (i.e., by classical mechanics).

QUANTIZATION OF ELECTRONIC ANGULAR MOMENTUM

Rutherford proposed that electrons orbit about the nucleus of an atom. One problem with this model is that, classically, orbiting electrons experience a centripetal acceleration, and accelerating charges lose energy by radiating; a stable electronic orbit is classically forbidden. Bohr nevertheless assumed stable electronic orbits with the electronic angular momentum quantized as

$$l = mvr = n\hbar$$

Quantization of angular momentum means that the radius of the orbit and the energy will be quantized as well. Bohr assumed that the discrete lines seen in the spectrum of the hydrogen atom were due to transitions of an electron from one allowed orbit/energy to another. He further assumed that the energy for a transition is acquired or released in the form of a photon as proposed by Einstein, so that

$$\Delta E = h\nu$$

This is known as the *Bohr frequency condition*. This condition, along with Bohr's expression for the allowed energy levels, gives a good match to the observed hydrogen atom spectrum. However, it works only for atoms with one electron.

WAVE-PARTICLE DUALITY

Einstein had shown that the momentum of a photon is

$$p = \frac{h}{\lambda}$$

This can be easily shown as follows. Assuming $E = hv$ for a photon and $\lambda v = c$ for an electromagnetic wave, we obtain

$$E = \frac{hc}{\lambda}$$

Now we use Einstein's relativity result $E = mc^2$ to find

$$\lambda = \frac{h}{mc}$$

which is equivalent to equation. Note that m refers to the relativistic mass, not the rest mass, since the rest mass of a photon is zero. Since light can behave both as a wave (it can be diffracted, and it has a wavelength), and as a particle (it contains packets of energy hv), de Broglie reasoned in 1924 that matter also can exhibit this *wave-particle duality*. He further reasoned that matter would obey the same equation as light. In 1927, Davisson and Germer observed diffraction patterns by bombarding metals with electrons, confirming de Broglie's proposition.

de Broglie's equation offers a justification for Bohr's assumption. If we think of an electron as a wave, then for the electron orbit to be stable the wave must complete an integral number of wavelengths during its orbit. Otherwise, it would interfere destructively with itself. This condition may be written as

$$2\pi r = n\lambda$$

If we use the de Broglie relation, this can be written as

$$mvr = n\hbar$$

which is identical to Bohr's equation.

Although de Broglie's equation justifies Bohr's quantization assumption, it also demonstrates a deficiency of

Bohr's model. Heisenberg showed that the wave-particle duality leads to the famous uncertainty principle

$$\Delta x \Delta p \approx h$$

One result of the uncertainty principle is that if the orbital radius of an electron in an atom r is known exactly, then the angular momentum must be completely unknown. The problem with Bohr's model is that it specifies r exactly and it also specifies that the orbital angular momentum must be an integral multiple of $\hbar$. Thus the stage was set for a new quantum theory which was consistent with the uncertainty principle.

THE SCHRODINGER EQUATION

In 1925, Erwin Schrodinger and Werner Heisenberg independently developed the new quantum theory. Schrodinger's method involves partial differential equations, whereas Heisenberg's method employs matrices; however, a year later the two methods were shown to be mathematically equivalent.

The Time-Independent Schrodinger Equation

Here we follow the treatment of McQuarrie, Section . We start with the one-dimensional classical wave equation,

$$\frac{\partial^2 u}{\partial x^2} = \frac{1}{v^2}\frac{\partial^2 u}{\partial t^2}$$

By introducing the separation of variables

$$u(x,t) = \psi(x) f(t)$$

we obtain

$$f(t)\frac{d^2\psi(x)}{dx^2} = \frac{1}{v^2}\psi(x)\frac{d^2 f(t)}{dt^2}$$

If we introduce one of the standard wave equation solutions for $f(t)$ such as $e^{i\omega t}$ (the constant can be taken care of later in the normalization), we obtain

$$\frac{d^2\psi(x)}{dx^2} = \frac{-\omega^2}{v^2}\psi(x)$$

Now we have an ordinary differential equation describing the spatial amplitude of the matter wave as a function of position. The energy of a particle is the sum of kinetic and potential parts

$$E = \frac{p^2}{2m} + V(x)$$

which can be solved for the momentum, p, to obtain

$$p = \{2m[E - V(x)]\}^{1/2}$$

Now we can use the de Broglie formula to get an expression for the wavelength

$$\lambda = \frac{h}{p} = \frac{h}{\{2m[E - V(x)]\}^{1/2}}$$

The term ω^2/v^2 in equation can be rewritten in terms of λ if we recall that $\omega = 2\pi v$ and $v\lambda = v$.

$$\frac{\omega^2}{v^2} = \frac{4\pi^2 v^2}{v^2} = \frac{4\pi^2}{\lambda^2} = \frac{2m[E - V(x)]}{\hbar^2}$$

When this result is substituted into equation we obtain the famous *time-independent Schrodinger equation*

$$\frac{d^2\psi(x)}{dx^2} + \frac{2m}{\hbar^2} = [E - V(x)]\psi(x) = 0$$

which is almost written in the form

$$-\frac{\hbar^2}{2m}\frac{d^2\psi(x)}{dx^2} + V(x)\psi(x) = E\psi(x)$$

This single-particle one-dimensional equation can easily be extended to the case of three dimensions, where it becomes

$$-\frac{\hbar^2}{2m}\nabla^2\psi(r) + V(r)\psi(r) = E\psi(r)$$

A two-body problem can also be treated by this equation if the mass m is replaced with a reduced mass μ.

It is important to point out that this analogy with the classical wave equation only goes so far. We cannot, for instance, derive the time-*dependent* Schrodinger equation in an analogous fashion (for instance, that equation involves the partial first derivative with respect to time instead of the partial second derivative). In fact, Schrodinger presented his time-independent equation first, and then went back and postulated the more general time-dependent equation.

The Time-Dependent Schrodinger Equation

We are now ready to consider the time-dependent Schrodinger equation. Although we were able to derive the single-particle time-independent Schrodinger equation starting from the classical wave equation and the de Broglie relation, the time-dependent Schrodinger equation cannot be derived using elementary methods and is generally given as a postulate of quantum mechanics. It is possible to show that the time-dependent equation is at least *reasonable* if not derivable, but the arguments are rather involved .

The single-particle three-dimensional time-dependent Schrodinger equation is

$$i\hbar\frac{\partial\psi(r,t)}{\partial t}=-\frac{\hbar^2}{2m}\nabla^2\psi(r,t)+V(r)\psi(r,t)$$

where V is assumed to be a real function and represents the potential energy of the system. *Wave Mechanics* is the branch of quantum mechanics with equation as its dynamical law. Note that equation does not yet account for spin or relativistic effects.

Of course the time-dependent equation can be used to derive the time-independent equation. If we write the wavefunction as a product of spatial and temporal terms, $\psi(r,t)=\psi(r)f(t)$, then equation becomes

$$\psi(r)i\hbar\frac{df(t)}{dt}=f(t)\left[-\frac{\hbar^2}{2m}\nabla^2+V(r)\right]\psi(r)$$

or

$$\frac{i\hbar}{f(t)}\frac{df}{dt}=\frac{1}{\psi(r)}\left[-\frac{\hbar^2}{2m}\nabla^2+V(r)\right]\psi(r)$$

Since the left-hand side is a function of *t* only and the right hand side is a function of *r* only, the two sides must equal a constant. If we tentatively designate this constant *E* (since the right-hand side clearly must have the dimensions of energy), then we extract two ordinary differential equations, namely

$$\frac{1}{f(t)}\frac{df(t)}{dt}=-\frac{iE}{\hbar}$$

and

$$-\frac{\hbar^2}{2m}\nabla^2\psi(r)+V(r)\psi(r)=E\psi(r)$$

The latter equation is once again the time-independent Schrodinger equation. The former equation is easily solved to yield

$$f(t)=e^{-eEt/\hbar}$$

The Hamiltonian in equation is a Hermitian operator, and the eigenvalues of a Hermitian operator must be real, so *E* is real. This means that the solutions *f*(*t*) are purely oscillatory, since *f*(*t*) never changes in magnitude (recall Euler's formula $e^{\pm i\theta}=\cos\theta\pm i\sin\theta$). Thus if

$$\psi(r,t)=\psi(r)e^{-iEt/\hbar}$$

then the total wave function $\psi(r,t)$ differs from $\psi(r)$ only by a phase factor of constant magnitude. There are some interesting consequences of this. First of all, the quantity $|\psi(r,t)|^2$ is time independent, as we can easily show:

$$|\psi(r,t)|^2=\psi^*(r,t)\psi(r,t)$$
$$=e^{iEt/\hbar}\psi^*(r)e^{-iEt/\hbar}\psi(r)=\psi^*(r)\psi(r)$$

Secondly, the expectation value for any time-independent operator is also time-independent, if $\psi(r,t)$ satisfies equation. By the same reasoning applied above,

$$<A>=\int \psi^*(r,t)\hat{A}\psi(r,t)=\int \psi^*(r)\hat{A}\psi(r)$$

For these reasons, wave functions of the form are called *stationary states*. The state $\psi(r,t)$ is "stationary," but the particle it describes is not!

Of course equation represents a particular solution to equation. The general solution to equation will be a linear combination of these particular solutions, i.e.

$$\psi(r,t)=\sum_i c_i e^{-iE_i t/\hbar}\psi_i(r)$$

MATHEMATICAL BACKGROUND

OPERATORS

Levine defines an *operator* as "a rule that transforms a given function into another function" (p. 33). The differentation operator *d*/*dx* is an example—it transforms a differentiable function *f*(*x*) into another function *f*′(*x*). Other examples include integration, the square root, and so forth. Numbers can also be considered as operators (they multiply a function). McQuarrie gives an even more general definition for an operator: "An *operator* is a symbol that tells you to do something with whatever follows the symbol" (p. 79). Perhaps this definition is more appropriate if we want to refer to the $\hat{C}_3$ operator acting on NH^3, for example.

Operators and Quantum Mechanics

In quantum mechanics, physical observables (e.g., energy, momentum, position, etc.) are represented mathematically by operators. For instance, the operator corresponding to energy is the Hamiltonian operator

$$\hat{H} = -\frac{\hbar^2}{2}\sum_i \frac{1}{m_i}\nabla_i^2 + V$$

where i is an index over all the particles of the system. We have already encountered the single-particle Hamiltonian in equation. The average value of an observable A represented by an operator $\hat{A}$ for a quantum molecular state $\psi(r)$ is given by the "expectation value" formula

$$< A >= \int \psi^*(r)\hat{A}\psi(r)dr$$

Basic Properties of Operators

Most of the properties of operators are obvious, but they are summarized below for completeness.

- The sum and difference of two operators $\hat{A}$ and $\hat{B}$ are given by

$$(\hat{A} + \hat{B})f = \hat{A}f + \hat{B}f$$

$$(\hat{A} - \hat{B})f = \hat{A}f - \hat{B}f$$

- The product of two operators is defined by

$$\hat{A}\hat{B}f \equiv \hat{A}[\hat{B}f]$$

- Two operators are equal if

$$\hat{A}f = \hat{B}f$$

for all functions

- The identity operator $\hat{1}$ does nothing (or multiplies by 1)

$$\hat{1}f = f$$

A common mathematical trick is to write this operator as a sum over a complete set of states (more on this later).

$$\sum_i |i\rangle\langle i| f = f$$

- The associative law holds for operators

  $\hat{A}(\hat{B}\hat{C}) = (\hat{A}\hat{B})\hat{C}$

- The commutative law does *not* generally hold for operators. In general, $\hat{A}\hat{B} \neq \hat{B}\hat{A}$. It is convenient to define the quantity

  $$[\hat{A},\hat{B}] \equiv \hat{A}\hat{B} - \hat{B}\hat{A}$$

which is called the commutator of $\hat{A}$ and $\hat{B}$. Note that the order matters, so that $[\hat{A},\hat{B}] = -[\hat{B},\hat{A}]$. If $\hat{A}$ and B happen to commute, then $[\hat{A},\hat{B}] = 0$.

- The n-th power of an operator $\hat{A}^n$ is defined as n successive applications of the operator, e.g.

  $\hat{A}^2 f = \hat{A}\hat{A}f$

- The exponential of an operator $e^{\hat{A}}$ is defined via the power series

  $$e^{\hat{A}} = \hat{1} + \hat{A} + \frac{\hat{A}^2}{2!} + \frac{\hat{A}^3}{3!} + \ldots$$

Linear Operators

Almost all operators encountered in quantum mechanics are *linear operators*. A linear operator is an operator which satisfies the following two conditions:

$$\hat{A}(f + g) = \hat{A}f + \hat{A}g$$

$$\hat{A}(cf) = c\hat{A}f$$

where c is a constant and f and g are functions. As an example, consider the operators d/dx and $()^2$. We can see that d/dx is a linear operator because

$$(d/dx)[f(x) + g(x)] = (d/dx)f(x) + (d/dx)g(x)$$

$$(d/dx)[cf(x)] = c(d/dx)f(x)$$

However, $()^2$ is not a linear operator because

$$(f(x) + g(x))^2 \neq (f(x))^2 + (g(x))^2$$

The only other category of operators relevant to quantum mechanics is the set of *antilinear* operators, for which

$$\hat{A}(\lambda f + \mu g) = \lambda^* \hat{A}f + \mu^* \hat{A}g$$

Time-reversal operators are antilinear.

Eigenfunctions and Eigenvalues

An *eigenfunction* of an operator $\hat{A}$ is a function f such that the application of $\hat{A}$ on f gives f again, times a constant.

$$\hat{A}f = kf$$

where k is a constant called the *eigenvalue*. It is easy to show that if $\hat{A}$ is a linear operator with an eigenfunction g, then any multiple of g is also an eigenfunction of $\hat{A}$.

When a system is in an *eigenstate* of observable A then the expectation value of A is the eigenvalue of the wavefunction. Thus if

$$\hat{A}\psi(r) = a\psi(r)$$

then

$$<A> = \int \psi^*(r)\hat{A}\psi(r)dr$$

$$= \int \psi^*(r)a\psi(r)dr$$

$$= a\int \psi^*(r)\psi(r)dr = a$$

assuming that the wavefunction is normalized to 1, as is generally the case. In the event that $\psi(r)$ is not or cannot be normalized (free particle, etc.) then we may use the formula

$$<A> = \frac{\int \psi^*(r)\hat{A}\psi(r)}{d} r \int \psi^*(r)\psi(r)dr$$

What if the wavefunction is a combination of eigenstates? Let us assume that we have a wavefunction which is a linear combination of two eigenstates of $\hat{A}$ with eigenvalues a and b.

$$\psi = c_a\psi_a + c_b\psi_b$$

where $\hat{A}\psi_a = a\psi_a$ and $\hat{A}\psi_b = b\psi_b$. Then what is the expectation value of A?

$$<A> = \underline{\int \psi^* \hat{A}\psi}$$

$$= \underline{\int [c_a\psi_a + c_b\psi_b]^* \hat{A}[c_a\psi_a + c_b\psi_b]}$$

$$= \underline{\int [c_a\psi_a + c_b\psi_b]^* [ac_a\psi_a + bc_b\psi_b]}$$

$$= \underline{a|c_a|^2 \int \psi_a^*\psi_a + bc_a^*c_b \int \psi_a^*\psi_b + ac_b^*c_a}$$

$$\underline{\int \psi_b^*\psi_a + b|c_b|^2 \int \psi_b^*\psi_b}$$

$$= a|c_a|2 + b|c_b|^2$$

assuming that ψ_a and ψ_b are orthonormal (shortly we will show that eigenvectors of Hermitian operators are orthogonal). Thus the average value of A is a weighted average of eigenvalues, with the weights being the squares of the coefficients of the eigenvectors in the overall wavefunction.

Hermitian Operators

As mentioned previously, the expectation value of an operator $\hat{A}$ is given by

$$<A> = \int \psi^*(r)\hat{A}\psi(r)dr$$

and all physical observables are represented by such expectation values. Obviously, the value of a physical observable such as energy or density must be real, so we require <*A*> to be real. This means that we must have <*A*> = <*A*>*, or

$$\int \psi^*(r)\hat{A}\psi(r)dr = \int (\hat{A}\psi(r))^* \psi(r)dr$$

Operators $\hat{A}$ which satisfy this condition are called *Hermitian*. One can also show that for a Hermitian operator,

$$\int \psi_1^*(r)\hat{A}\psi_2(r)dr = \int (\hat{A}\psi_1(r))^* \psi_2(r)dr$$

for any two states ψ_1 and ψ_2.

An important property of Hermitian operators is that their eigenvalues are real. We can see this as follows: if we have an eigenfunction of $\hat{A}$ with eigenvalue *a*, i.e. $\hat{A}\psi_a = a\psi_a$, then for a Hermitian operator $\hat{A}$

$$\underline{\int \psi_a^*\hat{A}\psi_a} = \underline{\int \psi_a(\hat{A}\psi_a)^*}$$

$$\underline{a\int \psi_a^*\psi_a} = \underline{a^*\int \psi_a\psi_a^*}$$

$$\underline{(a-a^*)\int |\psi_a|^2 = 0}$$

Since $|\psi_a|^2$ is never negative, we must have either $a = a^*$ or $\psi_a = 0$. Since $\psi_a = 0$ is not an acceptable wavefunction, $a = a^*$, so *a* is real.

Another important property of Hermitian operators is that their eigenvectors are orthogonal (or can be chosen to be so). Suppose that ψ_a and ψ_b are eigenfunctions of $\hat{A}$ with eigenvalues *a* and *b*, with $a \neq b$. If $\hat{A}$ is Hermitian then

$$\underline{\int \psi_a^*\hat{A}\psi_b} = \underline{\int \psi_b(\hat{A}\psi_a)^*}$$

$$b\int \psi_a^* \psi_b = a^* \int \psi_b \psi_a^*$$

$$(b-a)\int \psi_a^* \psi_b = 0$$

since $a = a^*$ as shown above. Because we assumed $b \neq a$, we must have $\int \psi_a^* \psi_b = 0$, i.e. ψ_a and ψ_b are orthogonal. Thus we have shown that eigenfunctions of a Hermitian operator with different eigenvalues are orthogonal. In the case of degeneracy (more than one eigenfunction with the same eigenvalue), we can *choose* the eigenfunctions to be orthogonal. We can easily show this for the case of two eigenfunctions of $\hat{A}$ with the same eigenvalue. Suppose we have

$$\hat{A}\psi_j = j\psi_j$$

$$\hat{A}\psi_k = j\psi_k$$

We now want to take linear combinations of ψ_j and ψ_k to form two new eigenfunctions $\psi_{j'}$ and $\psi_{k'}$, where $\psi_{j'} = \psi_j$ and $\psi_{k'} = \psi_k + c\psi_j$. Now we want $\psi_{j'}$ and $\psi_{k'}$ to be orthogonal, so

$$\int \psi_{j'}^* \psi_{k'} = 0$$

$$\int \psi_j^* (\psi_k + c\psi_j) = 0$$

$$\int \psi_j^* \psi_k + c\int \psi_j^* \psi_j = 0$$

Thus we merely need to choose

$$c = -\frac{\int \psi_j^* \psi_k}{\int \psi_j^* \psi_j}$$

and we obtain orthogonal eigenfunctions. This Schmidt-orthogonalization procedure can be extended to the case of nfold degeneracy, so we have shown that for a Hermitian operator, the eigenvectors can be made orthogonal.

Unitary Operators

A linear operator whose inverse is its adjoint is called *unitary*. These operators can be thought of as generalizations of complex numbers whose absolue value is 1.

$$U^{-1} = U^{t}$$

$$UU^{t} = U^{t}U = I$$

A unitary operator preserves the "lengths" and "angles" between vectors, and it can be considered as a type of rotation operator in abstract vector space. Like Hermitian operators, the eigenvectors of a unitary matrix are orthogonal. However, its eigenvalues are not necessarily real.

COMMUTATORS IN QUANTUM MECHANICS

The *commutator*, defined in section, is very important in quantum mechanics. Since a definite value of observable A can be assigned to a system only if the system is in an eigenstate of $\hat{A}$, then we can simultaneously assign definite values to two observables A and B only if the system is in an eigenstate of both $\hat{A}$ and $\hat{B}$. Suppose the system has a value of A_i for observable A and B_j for observable B. The we require

$$\hat{A}\psi_{A_i},B_j = B_i\psi_{A_i}B_j$$

$$\hat{B}\psi_{A_i},B_j = B_j\psi_{A_i}B_j$$

If we multiply the first equation by $\hat{B}$ and the second by $\hat{A}$ then we obtain

$$\hat{B}\hat{A}\psi_{A_i},B_j = \hat{B}\hat{A}_i\psi_{A_i}B_j$$

$$\hat{A}\hat{B}\psi_{A_i},B_j = \hat{A}\hat{B}_i\psi_{A_i}B_j$$

and, using the fact that ψ_{A_i},B_j is an eigenfunction of $\hat{A}$ and $\hat{B}$, this becomes

$$\hat{B}\hat{A}\psi_{A_i},B_j = A_iB_j\psi_{A_i}B_j$$

$$\hat{A}\hat{B}\psi_{A_i},B_j = B_jA_i\psi_{A_iB_j}$$

so that if we subtract the first equation from the second, we obtain

$$(\hat{A}\hat{B} - \hat{B}\hat{A})\psi_{A_iB_j} = 0$$

For this to hold for general eigenfunctions, we must have $\hat{A}\hat{B} = \hat{B}\hat{A}$, or $[\hat{A}\hat{B}] = 0$. That is, for two physical quantities to be simultaneously observable, their operator representations must commute.

Section 8.8 of Merzbacher contains some useful rules for evaluating commutators. They are summarized below.

$$[\hat{A},\hat{B}] + [\hat{B},\hat{A}] = 0$$

$$[\hat{A},\hat{A}] = 0$$

$$[\hat{A},\hat{B}+\hat{C}] = [\hat{A},\hat{B}] + [\hat{A},\hat{C}]$$

$$[\hat{A}+\hat{B},\hat{C}] = [\hat{A},\hat{C}] + [\hat{B},\hat{C}]$$

$$[\hat{A},\hat{B}\hat{C}] = [\hat{A},\hat{B}]\hat{C} + \hat{B}[\hat{A},\hat{C}]$$

$$[\hat{A}\hat{B},\hat{C}] = [\hat{A},\hat{C}]\hat{B} + \hat{A}[\hat{B},\hat{C}]$$

$$[\hat{A},[\hat{B},\hat{C}]] + [\hat{C},[\hat{A},\hat{B}]] + [\hat{B},[\hat{C},\hat{A}]] = 0$$

If $\hat{A}$ and $\hat{B}$ are two operators which commute with their commutator, then

$$[\hat{A},\hat{B}^n] = n\hat{B}^{n-1}[\hat{A},\hat{B}]] + [\hat{B},[\hat{C},\hat{A}]] = 0$$

If $\hat{A}$ and $\hat{B}$ are two operators which commute with their commutator, then

$$[\hat{A},\hat{B}^n] = n\hat{B}^{n-1}[\hat{A},\hat{B}]$$

$$[\hat{A}^n,\hat{B}] = n\hat{A}^{n-1}[\hat{A},\hat{B}]$$

We also have the identity (useful for coupled-cluster theory)

$$e^{\hat{A}}\hat{B}e^{-\hat{A}} = \hat{B} + [\hat{A},\hat{B}] + \frac{1}{2!}[\hat{A},[\hat{A},\hat{B}]] + \frac{1}{3!}[\hat{A},[\hat{A},[\hat{A},\hat{B}]]] + \dots$$

Finally, if $[\hat{A},\hat{B}] = i\hat{C}$ then the uncertainties in A and B, defined as $\Delta A^2 = \langle A^2 \rangle - \langle A \rangle^2$, obey the relation

$$(\Delta A)(\Delta B) \geq \frac{1}{2} |\langle C \rangle|$$

This is the famous Heisenberg uncertainty principle. It is easy to derive the well-known relation

$$(\Delta x)(\Delta p_x) \geq \frac{\hbar}{2}$$

from this generalized rule.

LINEAR VECTOR SPACES IN QUANTUM MECHANICS

We have observed that most operators in quantum mechanics are linear operators. This is fortunate because it allows us to represent quantum mechanical operators as matrices and wavefunctions as vectors in some linear vector space. Since computers are particularly good at performing operations common in linear algebra (multiplication of a matrix times a vector, etc.), this is quite advantageous from a practical standpoint.

In an n-dimensional space we may expand any vector Ψ as a linear combination of basis vectors

$$\Psi = \sum_{i=1}^{n} a_i \Psi_i$$

For a general vector space, the coefficients a_i may be complex; thus one should not be too quick to draw parallels to the expansion of vectors in three-dimensional Euclidean space. The coefficients a_i are referred to as the "components" of the state vector Ψ, and for a given basis, the components of a vector specify it completely. The components of the sum of two vectors are the sums of the components. If $\Psi = \sum a_i \Psi_i$ and $\Psi_b = \sum b_i \Psi_i$ then

$$\Psi_a + \Psi_b = \sum_i (a_i + b_i)\Psi_i$$

and similarly

$$\lambda \Psi_a = \sum_i (\lambda a_i)\Psi_i$$

The *scalar product* of two vectors is a complex number denoted by

$$(\Psi_b, \Psi_a) = (\Psi_a, \Psi_b)^*$$

where we have used the standard linear-algebra notation. If we also require that

$$(\Psi_a, \lambda \Psi_b) = \lambda^* (\Psi_a, \Psi_b)$$

then it follows that

$$(\lambda \Psi_a, \Psi_b) = \lambda^* (\Psi_a, \Psi_b)$$

We also require that

$$(\Psi_a, \Psi_b + \Psi_c) = (\Psi_a, \Psi_b) + (\Psi_a + \Psi_c)$$

If the scalar product vanishes (and if neither vector in the product is the null vector) then the two vectors are orthogonal.

Generally the basis is chosen to be orthonormal, such that

$$(\hat{\Psi}_i, \hat{\Psi}_j) = \delta_{ij}$$

In this case, we can write the scalar product of two arbitrary vectors as

$$(\Psi_a, \Psi_b) = \left(\sum_i a_i \hat{\Psi}_i, \sum_j b_j \hat{\Psi}_j \right)$$

$$= \sum_i \sum_j a_i^* b_j (\hat{\Psi}_i, \hat{\Psi}_j)$$

$$= \sum_i a_i^* b_i$$

This can also be written in vector notation as

$$(\Psi_a, \Psi_b) = (a_1^* a_2^* ... a_n^*) \begin{pmatrix} b_1 \\ b_2 \\ \vdots \\ b_n \end{pmatrix}.$$

It is useful at this point to introduce Dirac's bra-ket notation. We define a "bra" as

$$\langle \Psi_a | = (a_1^* a_2^* ... a_n^*)$$

and a "ket" as

$$| \Psi_a \rangle = \begin{pmatrix} a_1 \\ a_2 \\ \vdots \\ a_n \end{pmatrix}$$

A bra to the left of a ket implies a scalar product, so

$$\langle \Psi_a | \Psi_b \rangle = (\Psi_a, \Psi_b)$$

Sometimes in superficial treatments of Dirac notation, the symbol $\langle \Psi_a | \Psi_b \rangle$ is defined alternatively as

$$\langle \Psi_a | \Psi_b \rangle = \int \Psi_a^*(x) \Psi_b(x) dx$$

This is equivalent to the above definition if we make the connections $a_i = \Psi_a(x)$ and $b_i = \Psi_b(x)$. This means that our basis vectors are *every possible value of* x. Since x is continuous, the sum is replaced by an integral. Often only the subscript of the vector is used to denote a bra or ket; we may have written the above equation as

$$\langle a \mid b\rangle = \int \Psi_a^*(x)\Psi_b(x)dx$$

Now we turn our attention to matrix representations of operators. An operator $\hat{A}$ can be characterized by its effect on the basis vectors. The action of $\hat{A}$ on a basis vector $\hat{\Psi}_j$ yields some new vector Ψ_j' which can be expanded in terms of the basis vectors so long as we have a complete basis set.

$$\hat{A}\hat{\Psi}_j = \Psi_j' = \sum_i^n \hat{\Psi}_i A_{ij}$$

If we know the effect of $\hat{A}$ on the basis vectors, then we know the effect of $\hat{A}$ on any arbitrary vector because of the linearity of $\hat{A}$.

$$\Psi_b = \hat{A}\Psi_a = \hat{A}\sum_j a_j\hat{\Psi}_j = \sum_j a_j\hat{A}\hat{\Psi}_j = \sum_j\sum_i a_j\hat{\Psi}_i A_{ij}$$

$$= \sum_j \hat{\Psi}_i\left(\sum_j \hat{A}_{ij}a_j\right)$$

or

$$b_i = \sum_j A_{ij}a_j$$

This may be written in matrix notation as

$$\begin{pmatrix} b_1 \\ b_2 \\ \vdots \\ b_n \end{pmatrix} = \begin{pmatrix} A_{11} & A_{12} & \cdots & A_{1n} \\ A_{21} & A_{22} & \cdots & A_{2n} \\ \vdots & \vdots & & \vdots \\ A_{n1} & A_{n2} & \cdots & A_{nn} \end{pmatrix}\begin{pmatrix} a_1 \\ a_2 \\ \vdots \\ a_n \end{pmatrix}$$

We can obtain the coefficients A_{ij} by taking the inner product of both sides of equation with $\hat{\Psi}_i$, yielding

$$(\hat{\Psi}_i, \hat{A}\hat{\Psi}_j) = \left(\hat{\Psi}_i, \sum_k^n \hat{\Psi}_k A_{kj}\right)$$

$$= \sum_{k}^{n} A_{kj}(\hat{\Psi}_i, \hat{\Psi}_k)$$

$$= A_{ij}$$

since $(\hat{\Psi}_i, \hat{\Psi}_k) = \delta_{ik}$ due to the orthonormality of the basis. In bra-ket notation, we may write

$$A_{ij} = \langle i \,|\, \hat{A} \,|\, j \rangle$$

where i and j denote two basis vectors. This use of braket notation is consistent with its earlier use if we realise that $\hat{A} \,|\, j\rangle$ is just another vector $|\, j'\rangle$.

It is easy to show that for a linear operator $\hat{A}$, the inner product $(\Psi_a, \hat{A}\Psi_b)$ for two general vectors (not necessarily basis vectors) Ψ_a and Ψ_b is given by

$$(\Psi_a, \hat{A}\Psi_b) = \sum_{i}\sum_{j} a_i^* A_{ij} b_j$$

or in matrix notation

$$(\Psi_a, \hat{A}\Psi_b) = (a_1^* a_2^* \ldots a_n^*) \begin{pmatrix} A_{11} & A_{12} & \ldots & A_{1n} \\ A_{21} & A_{22} & \ldots & A_{2n} \\ \vdots & \vdots & & \vdots \\ A_{n1} & A_{n2} & \ldots & A_{nn} \end{pmatrix} \begin{pmatrix} b_1 \\ b_2 \\ \vdots \\ b_n \end{pmatrix}$$

By analogy to equation, we may generally write this inner product in the form

$$(\Psi_a, \hat{A}\Psi_b) = \langle a \,|\, \hat{A} \,|\, b \rangle \int \Psi_a^*(x) \hat{A} \Psi_b(x) dx$$

Previously, we noted that $(\Psi_a, \Psi_b) = (\Psi_b, \Psi_b)^*$, or $\langle a \,|\, b \rangle = \langle b \,|\, a \rangle^*$. Thus we can see also that

$$(\Psi_a, \hat{A}\Psi_b) = (\hat{A}\Psi_b, \Psi_a)^*$$

We now define the *adjoint* of an operator $\hat{A}$, denoted by $\hat{A}^t$, as that linear operator for which

$$(\Psi_a, \hat{A}\Psi_b) = (\hat{A}^t\Psi_a, \Psi_b)$$

That is, we can make an operator act *backwards* into "bra" space if we take it's adjoint.

With this definition, we can further see that

$$(\Psi_a, \hat{A}\Psi_b) = (\hat{A}\Psi_b, \Psi_a)^* = (\Psi_b, \hat{A}^t\Psi_a)^* = (\hat{A}^t\Psi_a, \Psi_b)$$

or, in braket notation,

$$\langle a | \hat{A} | b \rangle = \langle \hat{A}b | a \rangle^* = \langle b | \hat{A}^t | a \rangle^* = \langle \hat{A}^t a | b \rangle$$

If we pick $\Psi_a, \hat{\Psi}_i$ and $\Psi_b, \hat{\Psi}_j$ (i.e., if we pick two basis vectors), then we obtain

$$(\hat{A}\hat{\Psi}_i, \hat{\Psi}_j) = (\hat{\Psi}_i, \hat{A}^t\hat{\Psi}_j)$$

$$(\hat{\Psi}_j, \hat{A}\hat{\Psi}_i)^* = (\hat{\Psi}_i, \hat{A}^t\hat{\Psi}_j)$$

$$A_{ji}^* = \underline{A_{ij}^t}$$

But this is precisely the condition for the elements of a matrix and its adjoint! Thus the adjoint of the matrix representation of $\hat{A}$ is the same as the matrix representation of $\hat{A}^t$.

This correspondence between operators and their matrix representations goes quite far, although of course the specific matrix representation depends on the choice of basis. For instance, we know from linear algebra that if a matrix and its adjoint are the same, then the matrix is called Hermitian. The same is true of the operators; if

$$\hat{A} = \hat{A}^t$$

then $\hat{A}$ is a Hermitian operator, and all of the special properties of Hermitian operators apply to $\hat{A}$ or its matrix representation.

POSTULATES OF QUANTUM MECHANICS

In this section, we will present six postulates of quantum mechanics. Again, we follow the presentation of McQuarrie, with the exception of postulate, which McQuarrie does not include. A few of the postulates have already been discussed in section.

Postulate: The state of a quantum mechanical system is completely specified by a function $\psi(r, t)$ that depends on the coordinates of the particle(s) and on time. This function, called the wave function or state function, has the important property that $\Psi^*(r,t)\Psi(r,t)d\tau$ is the probability that the particle lies in the volume element $d\tau$ located at r at time t.

The wavefunction must satisfy certain mathematical conditions because of this probabilistic interpretation. For the case of a single particle, the probability of finding it *somewhere* is 1, so that we have the normalization condition

$$\int_{-\infty}^{\infty} \Psi^*(r,t)\Psi(r,t)d\tau = 1$$

It is customary to also normalize many-particle wavefunctions to 1. The wavefunction must also be single-valued, continuous, and finite.

Postulate: To every observable in classical mechanics there corresponds a linear, Hermitian operator in quantum mechanics.

This postulate comes about because of the considerations raised in section: if we require that the expectation value of an operator $\hat{A}$ is real, then $\hat{A}$ must be a Hermitian operator. Some common operators occuring in quantum mechanics are collected in Table.

Table: Physical observables and their corresponding quantum operators (single particle)

Observable Name	Observable Symbol	Operator Symbol	Operator
Position	r	$\hat{r}$	Multiply be r
Momentum	P	$\hat{p}$	$-i\hbar\left(\hat{i}\frac{\partial}{\partial x}+\hat{j}\frac{\partial}{\partial y}+\hat{k}\frac{\partial}{\partial z}\right)$
Kinetic energy	T	$\hat{T}$	$-\frac{\hbar^2}{2m}\left(\frac{\partial^2}{\partial x^2}+\frac{\partial^2}{\partial y^2}+\frac{\partial^2}{\partial z^2}\right)$
Potential energy	$V(r)$	$\hat{V}(r)$	Multiply by $V(r)$
Total energy	E	$\hat{H}$	$-\frac{\hbar^2}{2m}\left(\frac{\partial^2}{\partial x^2}+\frac{\partial^2}{\partial y^2}+\frac{\partial^2}{\partial z^2}\right)$ $+ V(r)$
Angular momentum	l_x	$\hat{l}_x$	$-i\hbar\left(y\frac{\partial}{\partial x}-z\frac{\partial}{\partial y}\right)$
			$-i\hbar\left(z\frac{\partial}{\partial x}-x\frac{\partial}{\partial z}\right)$
	l_z	$\hat{l}_z$	$\left(x\frac{\partial}{\partial y}-y\frac{\partial}{\partial x}\right)$

Postulate: In any measurement of the observable associated with operator $\hat{A}$, the only values that will ever be observed are the eigenvalues *a*, which satisfy the eigenvalue equation

$$\hat{A}\Psi = a\Psi$$

This postulate captures the central point of quantum mechanics — the values of dynamical variables can be quantized (although it is still possible to have a continuum of eigenvalues in the case of unbound states). If the system is in an eigenstate of $\hat{A}$ with eigenvalue *a*, then any measurement of the quantity *A* will yield *a*.

Although measurements must always yield an eigenvalue, the state does not have to be an eigenstate of $\hat{A}$

initially. An arbitrary state can be expanded in the complete set of eigenvectors of $\hat{A}$ ($\hat{A}\Psi_i = a_i\Psi_i$) as

$$\Psi = \sum_{i}^{n} c_i \Psi_i$$

where *n* may go to infinity. In this case we only know that the measurement of *A* will yield *one* of the values a_i, but we don't know which one. However, we do know the *probability* that eigenvalue a_i will occur—it is the absolute value squared of the coefficient, $|c_i|^2$, leading to the fourth postulate below.

An important second half of the third postulate is that, after measurement of ψ yields some eigenvalue a_i, the wavefunction immediately "collapses" into the corresponding eigenstate Ψ_i (in the case that a_i is degenerate, then ψ becomes the projection of ψ onto the degenerate subspace). Thus, measurement affects the state of the system. This fact is used in many elaborate experimental tests of quantum mechanics.

Postulate: If a system is in a state described by a normalized wave function ψ, then the average value of the observable corresponding to $\hat{A}$ is given by

$$<A> = \int_{-\infty}^{\infty} \Psi^* \hat{A} \Psi d\tau$$

Postulate: The wavefunction or state function of a system evolves in time according to the time-dependent Schrodinger equation

$$\hat{H}\Psi(r,t) = i\hbar \frac{\partial \Psi}{\partial t}$$

The central equation of quantum mechanics must be accepted as a postulate, as discussed in section.

Postulate: The total wavefunction must be antisymmetric with respect to the interchange of all coordinates of one fermion with those of another. Electronic spin must be included in this set of coordinates. The Pauli exclusion principle is a direct result of this *antisymmetry principle.*

We will later see that Slater determinants provide a convenient means of enforcing this property on electronic wavefunctions.

Some Analytically Soluble Problems

Quantum chemists are generally concerned with solving the time-independent Schrodinger equation. This equation can be solved analytically only in a few special cases. In this section we review the results of some of these analytically soluble problems.

The Particle in a Box

Consider a particle constrained to move in a single dimension, under the influence of a potential $V(x)$ which is zero for $0 \leq x \leq a$ and infinite elsewhere. Since the wavefunction is not allowed to become infinite, it must have a value of zero where $V(x)$ is infinite, so $\psi(x)$ is nonzero only within $[0, a]$. The Schrodinger equation is thus

$$-\frac{\hbar^2}{2m}\frac{d^2\psi}{dx^2} = E\psi(x) \qquad 0 \leq x \leq a$$

It is easy to show that the eigenvectors and eigenvalues of this problem are

$$\psi_n(x) = \sqrt{\frac{2}{a}}\sin\left(\frac{n\pi x}{a}\right) \quad 0 \leq x \leq a \;\; n = 1, 2, 3, \ldots$$

$$E_n = \frac{h^2 n^2}{8ma^2} \qquad n = 1, 2, \ldots$$

Extending the problem to three dimensions is rather straight forward;

The Harmonic Oscillator

Now consider a particle subject to a restoring force $F = -kx$, as might arise for a mass-spring system obeying Hooke's Law. The potential is then

$$V(x) = -\int_{-\infty}^{\infty}(-kx)dx$$

$$= V_0 + \frac{1}{2}kx^2$$

If we choose the energy scale such that $V_0 = 0$ then $V(x) = (1/2)kx^2$. This potential is also appropriate for describing the interaction of two masses connected by an ideal spring. In this case, we let x be the distance between the masses, and for the mass m we substitute the reduced mass μ. Thus the harmonic oscillator is the simplest model for the vibrational motion of the atoms in a diatomic molecule, if we consider the two atoms as point masses and the bond between them as a spring. The one-dimensional Schrodinger equation becomes

$$-\frac{\hbar^2}{2\mu}\frac{d^2\psi}{dx^2} + \frac{1}{2}kx^2\psi(x) = E\psi(x)$$

After some effort, the eigenfunctions are

$$\psi_n(x) = N_n H_n(\alpha^{1/2}x)e^{-\alpha x^2/2} \quad n = 0, 1, 2, \ldots$$

where H_n is the Hermite polynomial of degree n, and α and N_n are defined by

$$\alpha = \sqrt{\frac{k\mu}{\hbar^2}} \qquad N_n = \frac{1}{\sqrt{2^n n!}}\left(\frac{\alpha}{\pi}\right)^{1/4}$$

The eigenvalues are

$$E_n = \hbar\omega(n + 1/2)$$

with $\omega = \sqrt{k/\mu}$

The Rigid Rotor

The rigid rotor is a simple model of a rotating diatomic molecule. We consider the diatomic to consist of two point masses at a fixed internuclear distance. We then reduce the model to a one-dimensional system by considering the rigid rotor to have one mass fixed at the origin, which is orbited by the reduced mass μ, at a distance τ. The Schrodinger equation is

$$-\frac{\hbar^2}{2I}\left[\frac{1}{\sin\theta}\frac{\partial}{\partial\theta}\left(\sin\theta\frac{\partial}{\partial\theta}\right)+\frac{1}{\sin^2\theta}\frac{\partial}{\partial\phi^2}\right]\psi(r)=E\psi(r)$$

After a little effort, the eigenfunctions can be shown to be the spherical harmonics $Y_J^M(\theta,\phi)$, defined by

$$Y_J^M(\theta,\phi)=\left[\frac{(2J+1)}{4\pi}\frac{(J-|M|)!}{(J+|M|)!}\right]^{1/2}P_j^{|M|}(\cos\theta)e^{iM\phi}$$

where $P_J^{|M|}(x)$ are the associated Legendre functions. The eigenvalues are simply

$$E_J=\frac{\hbar^2}{2I}J(J+1)$$

Each energy level E_J is $2J + 1$-fold degenerate in M, since M can have values $-J, -J+1, ..., J-1, J$.

The Hydrogen Atom

Finally, consider the hydrogen atom as a proton fixed at the origin, orbited by an electron of reduced mass μ. The potential due to electrostatic attraction is

$$V(r)=-\frac{e^2}{4\pi\in_0 r}$$

In SI units. The kinetic energy term in the Hamiltonian is

$$\hat{T}=-\frac{\hbar^2}{2\mu}\nabla^2$$

so we write out the Schrodinger equation in spherical polar coordinates as

$$-\frac{\hbar^2}{2\mu}\left[\frac{1}{r^2}\frac{\partial}{\partial r}\left(r^2\frac{\partial\psi}{\partial r}\right)\frac{1}{r^2\sin\theta}\frac{\partial}{\partial\theta}\left(\sin\theta\frac{\partial\psi}{\partial\theta}\right)+\frac{1}{r^2\sin^2\theta}\frac{\partial^2\psi}{\partial\phi^2}\right]$$

$$-\frac{e^2}{4\pi\varepsilon_0 r}\psi(r,\theta,\phi) = E\psi(r,\theta,\phi)$$

It happens that we can factor $\psi(r,\theta,\phi)$ into $R(r)_{nl}Y_l^m(\theta,\phi)$, where $Y_l^m(\theta,\phi)$ are again the spherical harmonics. The radial part $R(r)$ then can be shown to obey the equation

$$-\frac{\hbar^2}{2\mu r^2}\frac{d}{dr}\left(r^2\frac{dR}{dr}\right)+\left[\frac{\hbar^2 l(l+1)}{2\mu r^2}+V(r)-E\right]R(r)=0$$

which is called the radial equation for the hydrogen atom. Its (messy) solutions are

$$R_{nl}(r) = -\left[\frac{(n-l-1)!}{2n[(n+l)!]^3}\right]^{1/3}\left(\frac{2}{na_0}\right)^{l+3/2} r^l e^{-r/na_0} L_{n+l}^{2l+1}\left(\frac{2r}{na_0}\right)$$

where $0 \le l \le n-1$, and a_0 is the Bohr radius, $\epsilon_0 h^2/\pi\mu e^2$.

The functions $L_{n+l}^{2l+1}(2r/na_0)$ are the associated Laguerre functions. The hydrogen atom eigenvalues are

$$E_n = -\frac{e^2}{8\pi\epsilon_0 a_0 n^2} \qquad n = 1, 2, \ldots$$

There are relatively few other interesting problems that can be solved analytically. For molecular systems, one must resort to approximate solutions.

Approximate Methods

The problems discussed in the previous section (harmonic oscillator, rigid rotator, etc.) are some of the few quantum mechanics problems which can be solved analytically. For the vast majority of chemical applications, the Schrodinger equation must be solved by approximate methods. The two primary approximation techniques are the variational method and perturbation theory.

Perturbation Theory

The basic idea of perturbation theory is very simple: we split the Hamiltonian into a piece we know how to solve (the "reference" or "unperturbed" Hamiltonian) and a piece we don't know how to solve (the "perturbation"). As long as the perburbation is small compared to the unperturbed Hamiltonian, perturbation theory tells us how to correct the solutions to the unperturbed problem to approximately account for the influence of the perturbation. For example, perturbation theory can be used to approximately solve an anharmonic oscillator problem with the Hamiltonian

$$\hat{H} = -\frac{\hbar^2}{2\mu}\frac{d^2}{dx^2} + \frac{1}{2}kx^2 + \frac{1}{6}\gamma x^3.$$

Here, since we know how to solve the harmonic oscillator problem, we make that part the unperturbed Hamiltonian (denoted $\hat{H}^{(0)}$), and the new, anharmonic term is the perturbation (denoted $\hat{H}^{(1)}$):

$$\hat{H}^{(0)} = -\frac{\hbar^2}{2\mu}\frac{d^2}{dx^2} + \frac{1}{2}kx^2,$$

$$\hat{H}^{(1)} = +\frac{1}{6}\gamma x^3.$$

Perturbation theory solves such a problem in two steps. First, obtain the eigenfunctions and eigenvalues of the unperturbed Hamiltonian, $\hat{H}^{(0)}$:

$$\hat{H}^{(0)}\Psi_n^{(0)} = E_n^{(0)}\Psi_n^{(0)}.$$

Second, correct these eigenvalues and/or eigenfunctions to account for the perturbation's influence. Perturbation theory gives these corrections as an infinite series of terms, which become smaller and smaller for well-behaved systems:

$$E_n = \underline{E_n^{(0)} + E_n^{(1)} + E_n^{(2)} + \ldots}$$

$$\Psi_n = \underline{\Psi_n^{(0)} + \Psi_n^{(1)} + \Psi_n^{(2)} + \ldots}$$

Quite frequently, the corrections are only taken through first or second order. According to perturbation theory, the first-order correction to the energy is

$$E_n^{(1)} = \int \Psi_n^{(0)*} \hat{H}^{(1)} \Psi_n^{(0)},$$

and the second-order correction is

$$E_n^{(2)} = \int \Psi_n^{(0)*} \hat{H}^{(1)} \Psi_n^{(1)}.$$

One can see that the first-order correction to the wavefunction, $\Psi_n^{(1)}$, seems to be needed to compute the second-order energy correction. However, it turns out that the correction $\Psi_n^{(1)}$ can be written in terms of the zeroth-order wavefunction as

$$\Psi_n^{(1)} = \sum_{i \neq n} \Psi_i^{(0)} \frac{\int \Psi_i^{(0)*} \hat{H}^{(1)} \Psi_n^{(0)}}{E_n^{(0)} - E_i^{(0)}}.$$

Substituting this in the expression for $E_n^{(2)}$, we obtain

$$E_n^{(2)} = \sum_{i \neq n} \frac{\left| \int \Psi_n^{(0)*} \hat{H}^{(1)} \Psi_i^{(0)} \right|^2}{E_n^{(0)} - E_i^{(0)}}.$$

Going back to the anharmonic oscillator example, the ground state wavefunction for the unperturbed problem is just

$$E_0^{(0)} = \frac{1}{2} \hbar\omega,$$

$$\Psi_0^{(0)}(x) = N_0 H_0(\alpha^{1/2} x) e^{-\alpha x^2/2}$$

$$= \left(\frac{\alpha}{\pi} \right)^{1/4} e^{-\alpha x^2/2}.$$

The first-order correction to the ground state energy would be

$$E_0^{(1)} = \left(\frac{\alpha}{\pi}\right)^{1/2} \int_{-\infty}^{\infty} \frac{1}{6}\gamma x^3 e^{-\alpha x^2}\, dx.$$

It turns out in this case that $E_0^{(1)} = 0$, since the integrand is odd. Does this mean that the anharmonic energy levels are the same as for the harmonic oscillator? No, because there are higher-order corrections such as $E_0^{(2)}$ which are not necessarily zero.

The Variational Method

The variational method is the other main approximate method used in quantum mechanics. Compared to perturbation theory, the variational method can be more robust in situations where it's hard to determine a good unperturbed Hamiltonian (i.e., one which makes the perturbation small but is still solvable). On the other hand, in cases where there is a good unperturbed Hamiltonian, perturbation theory can be more efficient than the variational method.

The basic idea of the variational method is to guess a "trial" wavefunction for the problem, which consists of some adjustable parameters called "variational parameters." These parameters are adjusted until the energy of the trial wavefunction is minimized. The resulting trial wavefunction and its corresponding energy are variational method approximations to the exact wavefunction and energy.

Why would it make sense that the best approximate trial wavefunction is the one with the lowest energy? This results from the Variational Theorem, which states that the energy of any trial wavefunction E is always an upper bound to the exact ground state energy ε_0. This can be proven easily. Let the trial wavefunction be denoted Φ. Any trial function can formally be expanded as a linear combination of the exact eigenfunctions Ψ_i. Of course, in practice, we don't know the Ψ_i, since we're assuming that we're applying the variational method to a problem we can't solve analytically. Nevertheless, that doesn't prevent us from using the exact eigenfunctions in our proof, since they certainly exist and form a complete set, even if we don't happen to know them. So, the trial wavefunction can be written

$$\Phi = \sum_i c_i \Psi_i,$$

and the approximate energy corresponding to this wavefunction is

$$E[\Phi] = \frac{\int \Phi^* \hat{H} \Phi}{\int \Phi^* \Phi}.$$

Substituting the expansion over the exaxt wavefunctions,

$$E[\Phi] = \frac{\sum_{ij} C_i^* C_j \int \Psi_i^* \hat{H} \Psi_j}{\sum_{ij} C_i^* C_j \int \Psi_i^* \Psi_j}.$$

since the functions Ψ_j are the exact eigenfunctions of $\hat{H}$, we can use $\hat{H}\Psi_j = \varepsilon_j \Psi_j$

$$E[\Phi] = \frac{\sum_{ij} C_i^* C_j \varepsilon_j \int \Psi_i^* \Psi_j}{\sum_{ij} C_i^* C_j \int \Psi_i^* \Psi_j}.$$

Now using the fact that eigenfunctions of a Hermitian operator form an orthonormal set (or can be made to do so),

$$E[\Phi] = \frac{\sum_i C_i^* C_i \varepsilon_j}{\sum_i C_i^* C_j}.$$

We now subtract the exact ground state energy ε_0 from both sides to obtain

$$E[\Phi]-\varepsilon_0=\frac{\sum_i C_i^* C_i(\varepsilon_j-\varepsilon_0)}{\sum_i C_i^* C_i}.$$

Since every term on the right-hand side is greater than or equal to zero, the left-hand side must also be greater than or equal to zero, or

$$E[\Phi]\geq\varepsilon_0 .$$

In other words, the energy of any approximate wavefunction is always greater than or equal to the exact ground state energy ε_0. This explains the strategy of the variational method: since the energy of any approximate trial function is always above the true energy, then any variations in the trial function which lower its energy are necessarily making the approximate energy closer to the exact answer. One example of the variational method would be using the Gaussian function $\phi(r)=e^{-\alpha r^2}$ as a trial function for the hydrogen atom ground state. This problem could be solved by the variational method by obtaining the energy of $\phi(r)$ as a function of the variational parameter α, and then minimizing $E(\alpha)$ to find the optimum value α_{min}. The variational theorem's approximate wavefunction and energy for the hydrogen atom would then be $\phi(r)=e^{-\alpha_{\text{min}} r^2}$ and $E(\alpha_{\text{min}})$.

Frequently, the trial function is written as a linear combination of basis functions, such as

$$\Phi=\sum_i c_i\phi_i .$$

This leads to the *linear variation method*, and the variational parameters are the expansion coefficients C_i. The energy for this approximate wavefunction is just

$$E[\Phi]=\frac{\sum_{ij} c_i^* c_j \int \phi_i^* \hat{H}\phi_j}{\sum_{ij} c_i^* c_j \int \phi_i^* \phi_j},$$

which can be simplified using the notation

$$H_{ij} = \int \phi_i^* \hat{H} \phi_j,$$

$$S_{ij} = \int \phi_i^* \phi_j,$$

to yield

$$E[\Phi] = \frac{\sum_{ij} c_i^* c_j H_{ij}}{\sum_{ij} c_i^* c_j S_{ij}}.$$

Differentiating this energy with respect to the expansion coefficients c_i yields a non-trivial solution only if the following "secular determinant" equals 0.

$$\begin{vmatrix} H_{11} - ES_{11} & H_{12} - ES_{12} & \cdots & H_{1N} - ES_{1N} \\ H_{21} - ES_{21} & H_{22} - ES_{22} & \cdots & H2N - ES_{2N} \\ \vdots & \vdots & \vdots & \vdots \\ H_{N1} - ES_{N1} & H_{N2} - ES_{N2} & \cdots & H_{NN} - ES_{NN} \end{vmatrix} = 0.$$

If an orthonormal basis is used, the secular equation is greatly simplified because S_{ij} is 1 for $i = j$ and 0 for $i \neq j$. In this case, the secular determinant is

$$\begin{vmatrix} H_{11} - E & H_{12} & \cdots & H_{1N} \\ H_{21} & H_{22} - E & \cdots & H_{2N} \\ \vdots & \vdots & \vdots & \vdots \\ H_{N1} & H_{N2} & \cdots & H_{NN} - E \end{vmatrix} = 0.$$

In either case, the secular determinant for N basis functions gives an N-th order polynomial in E which is solved for different roots, each of which approximates a different eigenvalue.

The variational method lies behind Hartree-Fock theory and the configuration interaction method for the electronic structure of atoms and molecules.

Molecular Quantum Mechanics

In this section, we discuss the quantum mechanics of atomic and molecular systems. We begin by writing the

Hamiltonian for a collection of nuclei and electrons, and then we introduce the Born-Oppenheimer approximation, which allows us to separate the nuclear and electronic degrees of freedom.

The Molecular Hamiltonian

We have noted before that the kinetic energy for a system of particles is

$$\hat{T} = -\frac{\hbar^2}{2}\sum_i \frac{1}{m_i}\nabla^2$$

The potential energy for a system of charged particles is

$$\hat{V}(r) = \sum_{i>j} \frac{Z_i Z_j e^2}{4\pi\epsilon_0} \frac{1}{|r_i - r_j|}$$

For a molecule, it is reasonable to split the kinetic energy into two summations—one over electrons, and one over nuclei. Similarly, we can split the potential energy into terms representing interactions between nuclei, between electrons, or between electrons and nuclei. Using i and j to index electrons, and A and B to index nuclei, we have (in atomic units)

$$\hat{H} = -\sum_A \frac{1}{2M_A}\nabla_A^2 - \sum_i \frac{1}{2}\nabla_i^2 + \sum_{A>B}\frac{Z_A Z_B}{R_{AB}} - \sum_{Ai}\frac{Z_A}{r_{Ai}} + \sum_{i>j}\frac{1}{r_{ij}}$$

where $r_{ij} = |r_i - r_j|$, $R_{Ai} = |r_A - r_i|$, and $R_{AB} = |r_A - r_B|$. This is known as the "exact" nonrelativistic Hamiltonian in field-free space. However, it is important to remember that this Hamiltonian neglects at least two effects. Firstly, although the speed of an electron in a hydrogen atom is less than 1% of the speed of light, relativistic mass corrections can become appreciable for the inner electrons of heavier atoms. Secondly, we have neglected the spin-orbit effects. From the point of view of an electron, it is being orbited by a nucleus which produces a magnetic field (proportional to L); this field interacts with the electron's magnetic moment (proportional to S), giving rise to a spin-orbit interaction (proportional to L.S for a diatomic.)

Although spin-orbit effects can be important, they are generally neglected in quantum chemical calculations.

The Born-Oppenheimer Approximation

We know that if a Hamiltonian is separable into two or more terms, then the total eigenfunctions are products of the individual eigenfunctions of the separated Hamiltonian terms, and the total eigenvalues are sums of individual eigenvalues of the separated Hamiltonian terms.

Consider, for example, a Hamiltonian which is separable into two terms, one involving coordinate q_1 and the other involving coordinate q_2.

$$\hat{H} = \hat{H}_1(q_1) + \hat{H}_2(q_2)$$

with the overall Schrodinger equation being

$$\hat{H}\psi(q_1, q_2) = E\psi(q_1, q_2)$$

If we assume that the total wavefunction can be written in the form $\psi(q_1, q_2) = \psi_1(q_1)\psi_2(q_2)$, where $\psi_1(q_1)$ and $\psi_2(q_2)$ are eigenfunctions of $\hat{H}_1$ and $\hat{H}_2$ with eigenvalues E_1 and E_2, then

$$\begin{aligned}
&\hat{H}\psi(q_1, q_2) = (\hat{H}_1 + \hat{H}_2)\psi_1(q_1)\psi_2(q_2)\\
&= \hat{H}_1\psi_1(q_1)\psi_2(q_2) + \hat{H}_2\psi_1(q_1)\psi_2(q_2)\\
&= E_1\psi_1(q_1)\psi_2(q_2) + E_2\psi_1(q_1)\psi_2(q_2)\\
&= (E_1 + E_2)\psi_1(q_1)\psi_2(q_2)\\
&= E\psi(q_1, q_2)
\end{aligned}$$

Thus the eigenfunctions of $\hat{H}$ are products of the eigenfunctions of $\hat{H}_1$ and $\hat{H}_2$, and the eigenvalues are the sums of eigenvalues of $\hat{H}_1$ and $\hat{H}_2$.

If we examine the nonrelativistic Hamiltonian, we see that the term

$$\sum_{Ai} \frac{Z_A}{r_{A_i}}$$

prevents us from cleanly separating the electronic and nuclear coordinates and writing the total wavefunction as $\psi(r,R) = \psi_e(r)\psi_N(R)$, where r represents the set of all electronic coordinates, and R represents the set of all nuclear coordinates. The Born-Oppenheimer approximation is to assume that this separation is nevertheless *approximately* correct.

Qualitatively, the Born-Oppenheimer approximation rests on the fact that the nuclei are much more massive than the electrons. This allows us to say that the nuclei are nearly fixed with respect to electron motion. We can fix R, the nuclear configuration, at some value R_a, and solve for $\psi_c(r;R_a)$; the electronic wavefunction depends only parametrically on R. If we do this for a range of R, we obtain the potential energy curve along which the nuclei move.

We now show the mathematical details. Let us abbreviate the molecular Hamiltonian as

$$\hat{H} = \hat{T}_N(R) + \hat{T}_c(r) + \hat{V}_{NN}(R) + \hat{V}_{cN}(r,R) + \hat{V}_{cc}(r)$$

where the meaning of the individual terms should be obvious. Initially, $\hat{T}_N(R)$ can be neglected since $\hat{T}_N$ is smaller than $\hat{T}_c$ by a factor of M_A/m_c, where m_c is the mass of an electron. Thus for a *fixed* nuclear configuration, we have

$$\hat{H}_{cl} = \hat{T}_c(r) + \hat{V}_{cN}(r;R) + \hat{V}_{NN}(R) + \hat{V}_{cc}(r)$$

such that

$$\hat{H}_{cl}\phi_c(r;R) = E_{cl}\phi_c(r;R)$$

This is the "clamped-nuclei" Schrodinger equation. Quite frequently $\hat{V}_{NN}(R)$ is neglected in the above equation, which is justified since in this case R is just a parameter so that $\hat{V}_{NN}(R)$ is just a constant and shifts the eigenvalues only by some constant amount. Leaving $\hat{V}_{NN}(R)$ out of the electronic Schrodinger equation leads to a similar equation,

$$\hat{H}_c = \hat{T}_c(r) + \hat{V}_{cN}(r;R) + \hat{V}_{cc}(r)$$

$$\hat{H}_c\phi_c(r;R) = E_c\phi_c(r;R)$$

where we have used a new subscript "e" on the electronic Hamiltonian and energy to distinguish from the case where $\hat{V}_{NN}$ is included.

We now consider again the original Hamiltonian. If we insert a wavefunction of the form $\phi_T(r,R) = \phi_c(r;R)\phi_N(R)$, we obtain

$$\hat{H}\phi_c(r;R)\phi_N(R) = E_{tot}\phi_c(r;R)\phi_c(r;R)\phi_N(R)$$

$$\{\hat{T}_N(R) + \hat{T}_c(r) + \hat{V}_{cN}(r,R) + \hat{V}_{NN}(R) + \hat{V}_{cc}(r)\phi_N(R)$$

$$= E_{tot}\phi_c(r;R)\phi_N(R)$$

Since $\hat{T}_e$ contains no R dependence,

$$\hat{T}_e\phi_e(r;R)\phi_N(R) = \phi_N(R)\hat{T}_e\phi_c(r;R)$$

However, we may not immediately assume

$$\hat{T}_N\phi_e(r;R)\phi_N(R) = \phi_e(r;R)\hat{T}_N\phi_N(R)$$

By the chain rule,

$$\nabla_A^2\phi_e(r;R)\phi_N(R) = \phi_e(r;R)\nabla_A^2\phi_N(R)$$

$$+2\nabla_A\phi_e(r;R)\nabla_A\phi_N(R) + \phi_N(R)\nabla_A^2\phi_e(r;R)$$

Using these facts, along with the electronic Schrodinger equation,

$$\{\hat{T}_e + \hat{V}_{cN}(r;R) + \hat{V}_{ee}(r)\}\phi_e(r;R) = \hat{H}_e\phi_e(r;R)E_e\phi_e(r;R)$$

we simplify (173) to

$$\phi_e(r;R)\hat{T}_N\phi_N(R)+\phi_N(R)\phi_c(r;R)E_e+\hat{V}_{NN})$$
$$-\left[\sum_A \frac{1}{2M_A}(2\nabla_A\phi_e(r;R)\nabla_A\phi_N(R)+\phi_N(R)\nabla_A^2\phi_e(r;R))\right]$$
$$=E_{tot}\phi_e(r;R)\phi_N(R)$$

We must now estimate the magnitude of the last term in brackets. Following Steinfeld, a typical contribution has the form $1/(2M_A)\nabla_A^2\phi_e(r;R)$, but $\nabla_A\phi_e(r;R)$ is of the same order as $\nabla_i\phi_e(r;R)$ since the derivatives operate over approximately the same dimensions. The latter is $\phi_e(r;R)p_e$, with p_e the momentum of an electron. Therefore $1/(2M_A)\nabla_A^2\phi_e(r;R)\ p_e^2/(2M_A)=(m/M_A)E_e$. Since $m/M_A \sim 1/10000$, the term in brackets can be dropped, giving

$$\phi_e(r;R)\hat{T}_N\phi_N(R)+\phi_N(R)E_E\phi_e(r;R)+\phi_N(R)\hat{V}_{NN}\phi_e(r;R)$$
$$E_{tot}\phi_e(r;R)\phi_N(R)$$
$$\{\hat{T}_N E_e+\hat{V}_{NN}\}\phi_N(R)=E_{tot}\phi_N(R)$$

This is the nuclear Shrodinger equation we anticipated—the nuclei move in a potential set up by the electrons.

To summarize, the large difference in the relative masses of the electrons and nuclei allows us to approximately separate the wavefunction as a product of nuclear and electronic terms. The electronic wavefucntion $\phi_e(r;R)$ is solved for a given set of nuclear coordinates,

$$\{\hat{H}_e\phi_e(r;R)=\left\{-\frac{1}{2}\sum_i\nabla_i^2-\sum_{A,i}\frac{Z_A}{r_{Ai}}+\sum_{A,i}\frac{1}{r_{ij}}\right\}$$
$$\phi_e(r;R)=E_e(r)\phi_e(r;R)$$

and the electronic energy obtained contributes a potential term to the motion of the nuclei described by the nuclear wavefunction $\phi_N(R)$.

$$\hat{H}_N \phi_N(R) = \left\{ -\sum_A \frac{1}{2M_A} \nabla_A^2 + E_e(R) + \sum_{A>B} \frac{Z_A Z_B}{R_{AB}} \right\} \phi_N(R)$$

$$= E_{tot} \phi_N(R)$$

Separation of the Nuclear Hamiltonian

The nuclear Schrodinger equation can be approximately factored into translational, rotational, and vibrational parts. McQuarrie explains how to do this for a diatomic in section. The rotational part can be cast into the form of the rigid rotor model, and the vibrational part can be written as a system of harmonic oscillators. Time does not allow further comment on the nuclear Schrodinger equation, although it is central to molecular spectroscopy.

Solving the Electronic Eigenvalue Problem

Once we have invoked the Born-Oppenheimer approximation, we attempt to solve the electronic Schrodinger equation, i.e.

$$\left[-\frac{1}{2} \sum_i \nabla_i^2 - \sum_{iA} \frac{Z_A}{r_{iA}} + \sum_{i>j} \frac{1}{r_{ij}} \right] \psi_c(r;R) = E_e \psi_e(r;R)$$

But, as mentioned previously, this equation is quite difficult to solve!

The Nature of Many-Electron Wavefunctions

Let us consider the nature of the electronic wavefunctions $\psi_e(r;R)$. Since the electronic wavefunction depends only parametrically on *R*, we will suppress *R* in our notation from now on. What do we require of $\psi_e(r)$? Recall that *r* represents the set of all electronic coordinates, i.e., r = $\{r_1, r_2, \dots r_N\}$. So far we have left out one important item—we need to include the *spin* of each electron. We can define a new variable *x* which represents the set of all four coordinates associated with an electron: three spatial coordinates r, and one spin coordinate ω, i.e., .

Thus we write the electronic wavefunction as $\psi_e (x_1, x_2, ..., x_N)$. Why have we been able to avoid including spin until now? Because the non-relativistic Hamiltonian does not include spin. Nevertheless, spin must be included so that the electronic wavefunction can satisfy a very important requirement, which is the *antisymmetry principle.* This principle states that for a system of fermions, the wavefunction must be antisymmetric with respect to the interchange of all (space *and* spin) coordinates of one fermion with those of another. That is,

$$\psi_e (x_1, ..., x_a, ..., x_b, ..., x_N) = -\psi_e (x_1, ..., x_b, ..., x_a, ..., x_N)$$

The Pauli exclusion principle is a direct consequence of the antisymmetry principle.

A very important step in simplifying $\psi_e (x)$ is to expand it in terms of a set of one-electron functions, or "orbitals." This makes the electronic Schrodinger equation considerably easier to deal with. A *spin orbital* is a function of the space and spin coordinates of a single electron, while a *spatial* orbital is a function of a single electron's spatial coordinates only. We can write a spin orbital as a product of a spatial orbital one of the two spin functions

$$\chi(x) = \psi(r)|\alpha\rangle$$

or

$$\chi(x) = \psi(r)|\beta\rangle$$

Note that for a given spatial orbital $\psi(r)$, we can form *two* spin orbitals, one with α spin, and one with β spin. The spatial orbital will be doubly occupied. It is possible (although sometimes frowned upon) to use one set of spatial orbitals for spin orbitals with α spin and another set for spin orbitals with β spin.

Where do we get the one-particle spatial orbitals $\psi(r)$? That is beyond the scope of the current section, but we briefly itemize some of the more common possibilities:

- Orbitals centered on each atom (atomic orbitals).
- Orbitals centered on each atom but also symmetry-adapted to have the correct point-group symmetry species (symmetry orbitals).
- Molecular orbitals obtained from a Hartree-Fock procedure.

We now explain how an *N*-electron function $\psi_e(x)$ can be constructed from spin orbitals, following the arguments of Szabo and Ostlund. Assume we have a complete set of functions of a single variable $\{\chi_i(x)\}$. Then any function of a single variable can be expanded exactly as

$$\Phi(x_1) = \sum_i a_i \chi_i(x_i) \; .$$

How can we expand a function of *two* variables, e.g. $\Phi(x_1, x_2)$?

If we hold x_2 fixed, then

$$\Phi(x_1, x_2) = \sum_i a_i(x_2) \chi_i(x_i)$$

Now note that each expansion coefficient $a_i(x_2)$ is a function of a single variable, which can be expanded as

$$a_i(x_2) = \sum_j b_{ij} \chi_j(x_2)$$

Substituting this expression into the one for $\Phi(x_1, x_2)$, we now have

$$\Phi(x_1, x_2) = \sum_{ij} b_{ij} \chi_j(x_1) \chi_j(x_2)$$

a process which can obviously be extended for $\Phi(x_1, x_2, ..., x_N)$.

We can extend these arguments to the case of having a complete set of functions of the variable x (recall x represents x, y, and z and also ω). In that case, we obtain an analogous result,

$$\Phi(x_1, x_2) = \sum_{ij} b_{ij} \chi_j(x_1) \chi_j(x_2)$$

Now we must make sure that the antisymmetry priciple is obeyed. For the two-particle case, the requirement

$$\Phi(x_1, x_2) = -\Phi(x_2, x_1)$$

implies that $b_{ij} = -b_{ji}$ and $b_{ii} = 0$, or

$$\Phi(x_1, x_2) = \sum_{j>i} b_{ij} [\chi_i(x_1) \chi_j(x_2) - \chi_j(x_1) \chi_i(x_2)]$$

$$\Phi(x_1, x_2) = \sum_{j>i} b_{ij} \, |\chi_i \chi_j\rangle$$

where we have used the symbol $\chi_i \chi_j\rangle$ to represent a Slater determinant, which in the general case is written

$$|\chi_1 \chi_2 \cdots \chi_N\rangle = \frac{1}{\sqrt{N!}} \begin{vmatrix} \chi_1(x_1) & \chi_2(x_1) & \cdots & \chi_N(x_1) \\ \chi_1(x_2) & \chi_2(x_2) & \cdots & \chi_N(x_2) \\ \vdots & \vdots & & \vdots \\ \chi_1(x_N) & \chi_2(x_N) & \cdots & \chi_N(x_N) \end{vmatrix}$$

We can extend the reasoning applied here to the case of N electrons; any N-electron wavefunction can be expressed exactly as a linear combination of all possible N-electron Slater determinants formed from a complete set of spin orbitals $\{\chi_i(x)\}$.

Index

H

I

J

K

L

M

N

P

Q

R

S

T

U

V

W

Z